GOVERNMENT PROCEDURES AND OPERATIONS

FEDERAL EMERGENCY MANAGEMENT

ELEMENTS AND CONSIDERATIONS

Government Procedures and Operations

Additional books in this series can be found on Nova's website under the Series tab.

Additional E-books in this series can be found on Nova's website under the E-book tab.

Defense, Security and Strategies

Additional books in this series can be found on Nova's website under the Series tab.

Additional E-books in this series can be found on Nova's website under the E-book tab.

GOVERNMENT PROCEDURES AND OPERATIONS

FEDERAL EMERGENCY MANAGEMENT

ELEMENTS AND CONSIDERATIONS

LEWIS B. ADAMS
EDITOR

nova publishers
New York

Library of Congress Cataloging-in-Publication Data

ISBN: 978-1-62618-217-2

Published by Nova Science Publishers, Inc. † New York

CONTENTS

Preface		**vii**
Chapter 1	Federal Emergency Management: A Brief Introduction *Bruce R. Lindsay*	**1**
Chapter 2	Presidential Policy Directive 8 and the National Preparedness System: Background and Issues for Congress *Jared T. Brown*	**37**
Chapter 3	Federal Stafford Act Disaster Assistance: Presidential Declarations, Eligible Activities, and Funding *Francis X. McCarthy*	**71**
Chapter 4	FEMA's Disaster Declaration Process: A Primer *Francis X. McCarthy*	**113**
Chapter 5	Federal Disaster Recovery Programs: Brief Summaries *Carolyn V. Torsell*	**145**
Index		**161**

PREFACE

The federal government plays a significant role in emergency management, which generally refers to activities associated with avoiding and responding to natural and human-caused hazards. Emergency management in the United States is highly decentralized and contextual in nature: activities often involve multiple jurisdictions as well as a vast number of agencies, nongovernmental organizations, and private sector entities. In addition, the number and type of actors involved in an incident will vary tremendously depending on the context and severity of the event. Similarly, the legal framework through which emergency management functions and activities are authorized is also decentralized and stems from multiple authorities. This book describes the process for requesting federal assistance for major disasters, emergencies, and fire suppression, with a focus on Presidential Policy Directive 8 and the National Preparedness System; the Federal Stafford Act Disaster Assistance; and federal disaster recovery programs.

Chapter 1 - The federal government plays a significant role in emergency management, which generally refers to activities associated with avoiding and responding to natural and human-caused hazards. Emergency management in the United States is highly decentralized and contextual in nature: activities often involve multiple jurisdictions as well as a vast number of agencies, nongovernmental organizations, and private sector entities. In addition, the number and type of actors involved in an incident will vary tremendously depending on the context and severity of the event. Similarly, the legal framework through which emergency management functions and activities are authorized is also decentralized and stems from multiple authorities. Congress annually appropriates funds for a wide range of activities and efforts related to emergency management. For example, between 2005 and 2011 Congress provided an average of $12 billion annually to the Federal Emergency Management Agency, the lead federal agency responsible for disaster relief through regular and supplemental appropriations. Congress has also invested over $120 billion through various federal agencies to help the Gulf Coast Region recover from the hurricanes that hit the Gulf Coast in 2005 and 2008. In recent years congressional interest in emergency management has focused on funding, program administration, and program coordination—both among federal agencies and state emergency management agencies. This report provides an introduction to the principles and foundations of federal emergency management in the United States and a description of the activities of the federal agencies that provide assistance, focusing primarily on the Federal Emergency Management Agency, but also including information on the National Guard, Department of Agriculture, Department of Defense, Army Corps of Engineers, Department of Health and Human Services, Department of Housing and Urban Development, Department of Transportation, Environmental Protection Agency, Forest Service, and Small Business Administration.

This report is designed to provide Members of Congress and congressional staff with a general overview of principles and foundations of federal emergency management in the United States as well as the types of activities provided by various federal agencies. The report begins with a description of the four phases of emergency management: (1) mitigation, (2) preparedness, (3) response, and (4) recovery, and includes examples of some of the activities that take place in each of these phases. The report then discusses a recent movement at the federal level to carry out these phases of emergency management through a system of frameworks. The frameworks include (1) the National Prevention Framework, (2) the National Protection Framework, (3) the National Mitigation Framework, (4) the National Response Framework, and (5) the National Disaster Recovery Framework. The frameworks are used to designate roles and responsibilities and coordinate various activities. Next, this report describes the process for requesting federal assistance for major disasters, emergencies, and fire suppression. The declaration section also includes brief summaries of the types of assistance provided through each type of declaration. This discussion is followed by description of federal-to-state cost shares, how federal assistance is funded, and the process through which FEMA requests assistance from other federal entities. The section then provides a description of the close-out process—the process in which FEMA terminates its recovery efforts. The report includes a discussion of key federal laws and policies that influence federal emergency management, and concludes by highlighting some of the federal activities that take place in response to emergencies and disasters.

Chapter 2 - Presidential Policy Directive 8: National Preparedness (PPD-8) was signed and released by President Barack Obama on March 30, 2011. PPD-8 and its component policies intend to guide how the nation, from the federal level to private citizens, can "prevent, protect against, mitigate the effects of, respond to, and recover from those threats that pose the greatest risk to the security of the Nation." These threats include terrorist acts, natural disasters, and other man-made incidents. PPD-8 evolves from, and supersedes, Homeland Security Presidential Directive 8, which was released under President George W. Bush. PPD-8 is intended to meet many requirements of Subtitle C of the Post-Katrina Emergency Reform Act of 2006 (P.L. 109-295, 6 U.S.C. §741-764). In addition to the main Directive, an Implementation Plan for PPD-8 and a National Preparedness Goal were finalized in 2011. Two National Planning Frameworks are also complete, but multiple component PPD-8 policy documents are still being developed. Some elements of PPD-8 may not be finalized until September 2012 or later. However, PPD-8 has already affected national preparedness policy by expanding the scope of the end-state objective for preparedness, modifying the capabilities-based planning methodology, identifying a new set of national capabilities, and directing the creation of more National Planning Frameworks. It is anticipated that the five National Planning Frameworks—one each for prevention, protection, mitigation, response, and recovery—will assign federal roles and responsibilities in each mission area. The National Planning Frameworks are also to guide how nonfederal resources are leveraged, including non-profit and private sectors' resources.

Congress may wish to oversee how the Administration creates and implements the many elements of PPD-8. This report discusses several potential issues and challenges that may arise in the development and implementation of each National Planning Framework. These issues and challenges include evaluating: how PPD-8 policies conform with statute; how federal roles and responsibilities have been assigned to implement and execute PPD-8 policies; how non-federal resources and stakeholders will be impacted by national preparedness guidance; and how the overall federal budget may be reprioritized by a new national preparedness goal. However, it may be difficult to ascertain the full impact of PPD-8 on national preparedness until its provisions are fully operationalized and tested during real world hazards.

Chapter 3 - The Robert T. Stafford Disaster Relief and Emergency Assistance Act (the Stafford Act) authorizes the President to issue major disaster or emergency declarations in response to catastrophes in the United States that overwhelm state and local governments. Such declarations result in the distribution of a wide range of federal aid to individuals and families, certain nonprofit organizations, and public agencies. Congress appropriates money to the Disaster Relief Fund (DRF), through both annual appropriations and emergency supplemental appropriations, for disaster assistance authorized by the Stafford Act. The Federal Emergency Management Agency (FEMA) within the Department of Homeland Security (DHS) administers most, but not all, of the authority the statute vests in the President.

The most recent significant action concerning the statute occurred in the closing months of the 109th Congress as a result of the congressional investigation on the response to Hurricane Katrina (August 2005). Senators inserted Stafford Act amendments into the FY2007 DHS appropriations legislation (Title VI of P.L. 109-295). These amendments expanded FEMA's authority to expedite emergency assistance to stricken areas, imposed new planning and preparedness requirements on federal administrators, provided new authority to regional offices, and increased federal assistance to victims and communities. More recently, Congress included a provision in the FY2010 appropriations legislation (P.L. 111-83) that allows retired law judges to arbitrate conflicts concerning the recovery of public infrastructure in the Gulf Coast due to Hurricanes Katrina and Rita. While not an amendment to the Stafford Act, this provision affects the administration of the FEMA appeals process under which applications for Stafford assistance are reconsidered. The decisions made to date by the arbitration panels resulted in an Administration request for supplemental funding that resulted in P.L. 111-112, which added more than $5.5 billion to the Disaster Relief Fund.

Previously introduced legislation during the 111th Congress would have amended the statute. Among the proposals, H.R. 3377, the Disaster Response, Recovery, and Mitigation Enhancement Act of 2009, would have authorized the Disaster Relief Fund, provided health benefits to temporary or intermittent federal employees who provide disaster assistance, authorized the National Urban Search and Rescue Response System, and requested FEMA to update standards for individual assistance disaster requests. Other bills sought to reauthorize a mortgage and rental assistance program terminated in 2000 (H.R. 888/S. 763), establish new eligibility criteria (H.R. 941, H.R. 1059, H.R. 1494, H.R. 2484, H.R. 4141, and S. 1069), and mandate establishment of a tracking and storage plan for housing units used by disaster survivors (H.R. 3437/S. 713).

Chapter 4 - The Robert T. Stafford Disaster Relief and Emergency Assistance Act (referred to as the Stafford Act - 42 U.S.C. 5721 et seq.) authorizes the President to issue "major disaster" or "emergency" declarations before or after catastrophes occur. Emergency declarations trigger aid that protects property, public health, and safety and lessens or averts the threat of an incident becoming a catastrophic event. A major disaster declaration, issued after catastrophes occur, constitutes broader authority for federal agencies to provide supplemental assistance to help state and local governments, families and individuals, and certain nonprofit organizations recover from the incident.

The end result of a presidential disaster declaration is well known, if not entirely understood. Various forms of assistance are provided, including aid to families and individuals for uninsured needs and assistance to state and local governments and certain non-profits in rebuilding or replacing damaged infrastructure. The amount of assistance provided through presidential disaster declarations has exceeded $140 billion. Often, in recent years, Congress has enacted supplemental appropriations legislation to cover unanticipated costs. While the amounts spent by the federal government on different programs may be reported, and the progress of the recovery can be observed, much less is known about the process that initiates all of this activity. Yet, it is a process that has resulted in an average of more than one disaster declaration a week over the last decade. The disaster declaration procedure is foremost a process that preserves the discretion of the governor to request assistance and the President to decide to grant, or not to grant, supplemental help. The process employs some measurable criteria in two broad areas: Individual Assistance that aids families and individuals and Public Assistance that is mainly for repairs to infrastructure. The criteria, however, also considers many other factors, in each category of assistance, that help decision makers assess the impact of an event on communities and states. Under current law, the decision to issue a declaration rests solely with the President. Congress has no formal role, but has taken actions to adjust the terms of the process. For example, P.L. 109-295 established an advocate to help small states with the declaration process. More recently, Congress introduced legislation, H.R. 3377, that would direct FEMA to update some of its criteria for considering Individual Assistance declarations. Congress continues to examine the process and has received some recommendations for improvements. Given the importance of the decision, and the size of the overall spending involved, hearings have been held to review the declaration process so as to ensure fairness and equity in the process and its results.

Chapter 5 - This report summarizes federal disaster assistance programs for use by Members of Congress and their staff as they address the needs of their states, communities, and constituents. A number of federal agencies provide assistance to individual victims, such as state, territorial, and local governments and non-governmental entities following a disaster. The federal forms of assistance include grants, loans, loan guarantees, temporary housing, and counseling. The programs summarized in this report fall into two broad categories. First are programs for disaster situations; most of these programs are administered by the Federal Emergency Management Agency (FEMA) and the Small Business Administration (SBA). Second are general assistance programs that may be used in either disaster situations, or to meet regular service needs. Many federal agencies, including the Departments of Health and Human Services, Housing and Urban Development, and Justice, administer programs that may be included in the second category. The programs in this report are primarily organized by recipient: individuals, state and local governments, or businesses. These programs address short-term needs such as food and shelter, and long-term needs such as repair of public utilities. This report includes a list of relevant federal agency websites that provide information on disaster responses, updates on recovery efforts, and resources on federal assistance programs.

In: Federal Emergency Management
Editor: Lewis B. Adams

ISBN: 978-1-62618-217-2

Chapter 1

FEDERAL EMERGENCY MANAGEMENT: A BRIEF INTRODUCTION*

Bruce R. Lindsay

SUMMARY

The federal government plays a significant role in emergency management, which generally refers to activities associated with avoiding and responding to natural and human-caused hazards. Emergency management in the United States is highly decentralized and contextual in nature: activities often involve multiple jurisdictions as well as a vast number of agencies, nongovernmental organizations, and private sector entities.

In addition, the number and type of actors involved in an incident will vary tremendously depending on the context and severity of the event. Similarly, the legal framework through which emergency management functions and activities are authorized is also decentralized and stems from multiple authorities.

Congress annually appropriates funds for a wide range of activities and efforts related to emergency management. For example, between 2005 and 2011 Congress provided an average of $12 billion annually to the Federal Emergency Management Agency, the lead federal agency responsible for disaster relief through regular and supplemental appropriations.

---

* This is an edited, reformatted and augmented version of Congressional Research Service, Publication No. R42845, dated November 30, 2012.

Congress has also invested over $120 billion through various federal agencies to help the Gulf Coast Region recover from the hurricanes that hit the Gulf Coast in 2005 and 2008.

In recent years congressional interest in emergency management has focused on funding, program administration, and program coordination—both among federal agencies and state emergency management agencies. This report provides an introduction to the principles and foundations of federal emergency management in the United States and a description of the activities of the federal agencies that provide assistance, focusing primarily on the Federal Emergency Management Agency, but also including information on the National Guard, Department of Agriculture, Department of Defense, Army Corps of Engineers, Department of Health and Human Services, Department of Housing and Urban Development, Department of Transportation, Environmental Protection Agency, Forest Service, and Small Business Administration.

This report is designed to provide Members of Congress and congressional staff with a general overview of principles and foundations of federal emergency management in the United States as well as the types of activities provided by various federal agencies.

The report begins with a description of the four phases of emergency management: (1) mitigation, (2) preparedness, (3) response, and (4) recovery, and includes examples of some of the activities that take place in each of these phases. The report then discusses a recent movement at the federal level to carry out these phases of emergency management through a system of frameworks.

The frameworks include (1) the National Prevention Framework, (2) the National Protection Framework, (3) the National Mitigation Framework, (4) the National Response Framework, and (5) the National Disaster Recovery Framework. The frameworks are used to designate roles and responsibilities and coordinate various activities.

Next, this report describes the process for requesting federal assistance for major disasters, emergencies, and fire suppression. The declaration section also includes brief summaries of the types of assistance provided through each type of declaration. This discussion is followed by description of federal-to-state cost shares, how federal assistance is funded, and the process through which FEMA requests assistance from other federal entities.

The section then provides a description of the close-out process—the process in which FEMA terminates its recovery efforts. The report includes a discussion of key federal laws and policies that influence federal emergency management, and concludes by highlighting some of the federal activities that take place in response to emergencies and disasters.

INTRODUCTION

Emergency management generally refers to activities associated with avoiding and responding to natural and human-caused hazards. Emergency management in the United States is highly decentralized and contextual in nature.

Multiple jurisdictions as well as a vast number of agencies, nongovernmental organizations, and private sector entities are often involved. In general, emergency management begins locally, but the federal government plays an important role when a state requests assistance.

Consequently, the number and type of actors involved in an incident vary tremendously depending on the context and severity of the event. Similarly, the legal framework through which emergency management functions and activities are authorized is also decentralized and stems from multiple authorities.

This report provides Members of Congress and their staffs with an introduction to the principles and foundations of federal emergency management in the United States. It examines the activities of several federal agencies including the Federal Emergency Management Agency (FEMA) the National Guard, Department of Agriculture, Department of Defense, Army Corps of Engineers, Department of Health and Human Services, Department of Housing and Urban Development, Department of Transportation, Environmental Protection Agency, Forest Service, and Small Business Administration.

In addition, this report discusses the four phases of emergency management: (1) mitigation, (2) preparedness, (3) response, and (4) recovery; the process for requesting federal assistance for major disasters, emergencies, and fires; and the types of assistance provided through each type of Stafford Act declaration.

This report also includes a description of federal-to-state cost shares under the Stafford Act, a discussion on how federal assistance is funded, and the process through which FEMA requests assistance from other federal entities.

This report also outlines the frameworks that guide various emergency management activities at the federal and state level, and discusses some of the key federal laws and policies influencing federal emergency management and highlights federal entities that provide assistance to states and localities.

Related CRS products examining these issues more in-depth are footnoted in this report.

Key Concepts and Approaches in Emergency Management

The following sections describe key concepts that undergird federal emergency management. Many of these concepts originated at the state level and are still being put to use by states and localities. Some concepts developed by the states have been modified and/or adopted by the federal government as a national standard.

All-Hazards Model

The all-hazards model is based on the idea that there are generic processes and capabilities needed to address most kinds of emergencies and disasters. For example, preparing and responding to an earthquake entails similar activities and capabilities for preparing and responding to an explosion or terrorist bombing. Thus emergency managers can conduct emergency management activities in a more flexible and cost-effective manner than using a standalone, emergency, or disaster-specific program. Another benefit is that carrying out emergency management functions generally involves adaptation, which is a guiding principle of the all-hazards model.[1] Thus preparations and lessons learned associated with one type of event, can often be applied to another type of emergency or disaster scenario.

NIMS and ICS

Emergency management functions are managed according to the principles of the National Incident Management System (NIMS). Authorized by Homeland Security Presidential Directive 5 (HSPD-5), NIMS is a preparedness and response management model based on the Incident Command System (ICS).[2] ICS is a command and control model developed by firefighters after the 1970 fires in southern California. The response to the fires was hindered due to duplication of efforts, lack of coordination, and communication problems. ICS standardizes response operations by using similar terminology, communication systems, and organizational structure to eliminate or reduce confusion during a unified response.[3] NIMS uses ICS concepts to establish a response structure that is scalable (capable of growing as more organizations come together to respond to the incident) that can be used by all jurisdictions, agencies, and organizations to ensure a unified response to complex events. State and local governments must be NIMS compliant to be eligible for certain preparedness grants.

Phases of Emergency Management

Emergency management functions are generally grouped into four phases: (1) Mitigation, (2) Preparedness, (3) Response, and (4) Recovery. The grouping of emergency management functions is useful for classifying and conceptualizing activities.

Use of the four phases at the state level is not, however, a requirement for grant funding. As discussed later, the federal government uses a framework approach that differs somewhat from the four phases. The following sections provide examples of the types of activities that take place in each phase.[4]

While conceptually useful for targeting efforts and resources, the phases of emergency management are not distinct—activities in each phase often overlap.

For example, recovery projects often include elements of mitigation (for example, rebuilding structures using current building codes) and response often includes recovery measures (immediate debris removal). The phases are also cyclical in nature—lessons learned from an incident might be applied in preparedness efforts for future emergencies and major disasters.

Mitigation

Mitigation activities entail identifying risks and hazards to either substantially reduce or eliminate the impact of an incident usually through structural measures.

Mitigation activities often have a long-term or sustained effect and may have an impact on insurance premiums. In many cases, mitigation activities occur in the recovery stage of a major disaster. Some examples of mitigation include:

- building codes that address risks such as fires, high winds, or earthquakes;
- zoning rules that restrict construction in floodplains;
- rebuilding damaged structures with more resilient materials;

- flood mapping to identify low lying areas and relocating homes and structures located in floodplains and flood prone areas; and
- dams and levees that help prevent flooding.

Preparedness

Preparedness is distinct from mitigation because rather than focusing on eliminating or reducing risks, the general focus of preparedness is to enhance the capacity to respond to an incident by taking steps to ensure personnel and entities are capable of responding to a wide range of potential incidents. Preparedness activities may include:

- training;
- planning;
- procuring resources, such as food, water, and medication stockpiles;
- intelligence and surveillance activities to identify potential threats; and
- exercising to assure the adequacy of planning efforts and the use of after-action reports to improve emergency response plans.

Response

Response activities are comprised of the immediate actions to save lives, protect property and the environment, and meet basic human needs. Response involves the execution of emergency plans and related actions, and may include:

- evacuating victims;
- deployment of response teams, medical stockpiles, and other assets; and
- establishment of incident command operations.

Recovery

Recovery activities are intended to restore essential services and repair damages caused by the event. Recovery activities may include:

- the reconstitution of government operations and services (e.g., emergency services, public safety, and schools);
- housing and services for displaced families and individuals; and
- replenishment of stockpiles.

The Framework Approach

On March 30, 2011, President Barack Obama issued Presidential Policy Directive 8: National Preparedness (hereafter PPD-8).[5] PPD-8 superseded Homeland Security Presidential Directive 8 (HSPD-8) which was signed by President George W. Bush on December 17, 2003.[6] Similar to its predecessor, PPD-8 provides a guide as to how the nation, from the federal level to private citizens, can "prevent, protect against, mitigate the effects of, respond to, and recover from those threats that pose the greatest risk to the security of the Nation" including acts of terrorism and other human caused incidents (such as oil spills) as natural disasters.[7] PPD-8 is also intended to meet several "comprehensive preparedness system" requirements in Subtitle C of the Post-Katrina Act.[8]

PPD-8 establishes that preparedness objectives are to be accomplished by subdividing emergency management components into a system of integrated, national planning frameworks according to functionality—prevention, protection, mitigation, response, and recovery.[9] The frameworks are intended to assign roles and responsibilities to various federal agencies with mission areas involved with aspects of federal emergency management.[10]

These frameworks either exist in various stages of development or have been implemented. The following sections provide brief descriptions of each framework.[11]

National Prevention Framework

While the other planning frameworks address natural hazards and disasters as well as human-caused incidents, the focus of the National Prevention Framework is mainly on preventing imminent terrorist threats. As such, upon issue, the National Prevention Framework is to assign roles and responsibilities and coordinate federal agencies involved with intelligence and information sharing, surveillance, providing public information and warnings, as well as other elements that help the federal government identify, discover, or locate terrorist threats.[12]

National Protection Framework

The National Protection Framework, upon issue, would assign roles and responsibilities and coordinate agencies on a wide range of emergency management and homeland security areas, encompassing cyber security, border security, transportation security, and agriculture and food security, among others. In addition, the working draft of the National Protection

Framework proposes the use of academic and research centers to develop new protection technologies and establish protection-related curricula and degree programs.[13]

National Mitigation Framework

The National Mitigation Framework, upon issue, would address capabilities that reduce the loss of life and property by lessening the impact of disasters.[14] The National Mitigation Framework is distinct from the other planning frameworks because most mitigation activities take place at the local level, the role of the federal government in the National Mitigation Framework is not as prominent as with the other planning frameworks.[15]

National Response Framework

Issued in January 2008, the National Response Framework (NRF) is the successor of two previous response documents, the Federal Response Plan, which was thought to be too narrow in scope, and the National Response Plan, which was found to be problematic for a variety of reasons, including unclear designations and confusing language and jargon.[16] The NRF guides the federal response to natural and human-caused incidents. However, the NRF is not an "operational plan." Rather, it articulates the overarching emergency management principles used to coordinate and conduct a multi-agency and multijurisdictional response to all types of incidents. The NRF is executed through the use of three supplemental annexes consisting of the (1) Emergency Support Functions Annex, (2) Support Annexes, and (3) Incident Annexes.

Emergency Support Functions Annexes

There are 15 Emergency Support Function (ESF) Annexes. ESFs group federal departments and agencies by matching their resources and capabilities with a particular incident.[17] For example, federal entities with a role in responding to an oil spill are listed in ESF #10—the Oil and Hazardous Materials Response Annex. ESFs also designate which federal entities have management oversight reasonability and which entities have a support role.

Support Annexes

There are eight Support Annexes that group federal, state, local, private sector, and nongovernmental organizations that execute functional processes and administrative functions. The Support Annexes also designate roles and responsibilities.[18]

Incident Annexes

There are eight Incident Annexes that explain the authorities and policies relevant to a particular incident, describe the incident situation, and make planning assumptions for the incident. An Incident Annex also identifies the "coordinating and cooperating" agencies involved with response to the incident.[19]

National Disaster Recovery Framework

Issued in September 2011, the National Disaster Recovery Framework (NDRF) coordinates and assigns roles and responsibilities to entities involved in disaster recovery. The focus of the NDRF is to "restore, redevelop and revitalize the health, social, economic, natural and environmental" aspects of disaster-impacted states and local jurisdictions.[20] The NDRF has six Recovery Support Functions (RSF): (1) Community Planning and Capacity Building, (2) Economic, (3) Health and Social Services, (4) Housing, (5) Infrastructure Systems, and (6) Natural and Cultural Resources.

Federal Assistance through Stafford Declarations

The system of emergency management in the United States is scalable. This means that local governments request assistance from the state if responding or recovering from the incident is beyond their capacity. In cases when a state is overwhelmed by the incident, the state governor may elect to request assistance from the federal government.

Scalability makes emergency management response more practical, but it also contains a political element because it is embedded within the federalist system of governance aimed at the preservation of state autonomy. The state-initiated request may also alleviate concern that the federal government might assume leadership of response and recovery operations. The Robert T. Stafford Disaster Relief and Emergency Assistance Act (hereinafter the Stafford Act)[21] does provide the President authority to issue an emergency declaration in the absence of a gubernatorial request if the President determines the incident involves a subject area under the Constitution or laws of the United States, in which the United States exercises preeminent responsibility and authority of the incident.[22] Such cases, however, are rare. The majority of federal disaster assistance is released only after a presidential declaration is issued in response to a gubernatorial request for federal assistance.

The Stafford Act authorizes federal assistance through three types of declarations: (1) major disaster declarations, (2) emergency declarations, and (3) Fire Management Assistance Grant declarations (FMAG).[23] However, FMAG declarations are typically declared through the Federal Emergency Management Agency (FEMA) Regional Director.[24] Emergency and major disaster declarations can only be issued by the President. Each of these declarations are described in more detail in the following sections.

Major Disaster Declarations

The Stafford Act defines a major disaster as:

> any natural catastrophe (including any hurricane, tornado, storm, high water, wind-driven water, tidal wave, tsunami, earthquake, volcanic eruption, landslide, mudslide, snowstorm, or drought), or, regardless of cause, any fire, flood, or explosion, in any part of the United States, which in the determination of the President causes damage of sufficient severity and magnitude to warrant major disaster assistance under this chapter to supplement the efforts and available resources of states, local governments, and disaster relief organizations in alleviating the damage, loss, hardship, or suffering caused thereby.[25]

When a governor submits a request for major disaster declaration FEMA meets with state representatives to develop preliminary damage assessments. In general, FEMA will make a recommendation to the President to declare a major disaster if the state's preliminary damage assessment exceeds certain thresholds established in regulation.[26] The President in turn may or may not act on the recommendation.

Assistance Provided Under Major Disaster Declarations

Under the Stafford Act three main types of assistance can be provided and administered through FEMA when the President issues a major disaster declaration: (1) the Public Assistance (PA) Grant Program, (2) the Individual Assistance (IA) program, and (3) the Hazard Mitigation Grant Program (HMGP). Under the Stafford Act, the President may issue a major disaster declaration that provides only PA if damage to dwellings is not severe enough to warrant IA, or if there is sufficient damage, both PA and IA may be provided.

The PA program provides assistance to state and local governments, and certain nonprofit organizations and includes emergency protective measures, debris removal, and the repair, replacement, or restoration of eligible facilities.[27] Hazard mitigation measures during the recovery process are also provided under the PA program. However, not all of the entities affected by the incident are eligible for assistance—aid is only provided to entities that have been determined by FEMA officials to be in need of the assistance. There is no cap to the amount of assistance that can be provided through the PA program. The IA program is also limited to those deemed to be in need of assistance by FEMA officials. IA assistance includes temporary housing, disaster unemployment, legal services, crisis counseling and other needs assistance (e.g., disaster-related medical, dental, and burial costs), and the repair and replacement of uninsured damages to a home. At the time of this writing, the maximum grant amount available to an individual household is roughly $31,400.[28] HMGP provides grants to states in which major disasters have been declared. These funds must be used for activities that prevent future disasters or reduce their impact if they cannot be prevented. The Post-Katrina Act[29] adjusted the percentage amounts for HMGP awards by establishing a scale that authorizes a higher percentage (15% of the total Stafford Act assistance in a state) for major disasters in which no more than $2 billion is provided, 10% for assistance that ranges from more than $2 billion to $10 billion, and 7.5% for major disasters that involve Stafford Act assistance of more than $10 billion to $35.3 billion.[30]

Emergency Declarations[31]

The Stafford Act defines an emergency as:

> any occasion or instance for which, in the determination of the President, federal assistance is needed to supplement State and local efforts and capabilities to save lives and to protect property and public health and safety, or to lessen or avert the threat of a catastrophe in any part of the United States.[32]

As with disaster declarations, FEMA generally meets with state representatives to make a recommendation to the President on whether to issue an emergency declaration. However, unlike major disasters, a per capita formula to assess the magnitude of the disaster's costs is not used for the recommendation.

Assistance Provided under Emergency Declarations

Emergency declarations authorize activities that can help states and localities carry out essential services and activities that may reduce future damage. As such, emergency declarations may be declared before an incident occurs to save lives and lessen the impact of the incident. For example, emergency declarations have been declared prior to a hurricane making landfall to help state and local governments conduct evacuations and pre-position resources.[33] The types of assistance provided after an emergency declaration is declared include debris removal, distribution of food and medicine, and technical advisory assistance to state and local governments. Emergency declarations do not provide assistance for repairs and replacement of public infrastructure or nonprofit facilities, nor do they provide assistance to individuals and households.

Fire Management Assistance Grant Declarations

Section 420 of the Stafford Act authorizes various forms of federal assistance, such as equipment, personnel, and grants to any state or local government for the control, management and mitigation of any fire on public or private forest land or grassland that might become a major disaster.[34] FEMA meets with state representatives when a governor submits a request for a FMAG, but does not use a per capita formula to assess the magnitude of the fire's costs when making recommendations whether to issue a FMAG declaration.

Assistance Provided under Fire Management Assistance Grants

Section 420(c) authorizes the President to provide essential assistance under Section 403 of the Stafford Act.[35] Essential assistance includes the distribution of medicine, medical equipment, and food through local governments, and disaster assistance organizations such as the National Red Cross and the Salvation Army. Essential assistance also includes debris removal, search and rescue, emergency mass care and shelter, clearance of roads and temporary structures such as bridges and facilities for schools, and other services deemed essential to the community. Other essential assistance includes warnings, and the dissemination of public information. FMAG does not provide long term assistance to individuals and households.

COST-SHARES

Under the Stafford Act, the costs of response and recovery efforts are usually split between the state and the federal government. The Stafford Act contains language detailing the amount the federal government must pay to the state for eligible repairs, restoration projects, and replacements of damaged facilities. The formula is called a "cost-share" and the Stafford Act provides that the federal share "shall be not less than 75 percent."[36] The cost-share can be adjusted to reduce, but not increase, the state share. The reduced state share is known as a cost-share waiver. In most cases, waivers are adjusted according to FEMA regulatory authority and carried out by the executive branch through administrative actions based on per capita preliminary damage estimates.[37] Additionally, since 1997, and particularly in the wake of concerns raised about state capacity to meet their cost-shares following the Gulf Coast storms of 2005, Congress has begun to exercise its authority to adjust cost-shares.[38]

DISASTER RELIEF FUND

The majority of assistance provided under the Stafford Act is paid out of FEMA's Disaster Relief Fund (DRF).[39] The DRF is a no-year account[40] used to fund response activities and pay for ongoing recovery programs. The DRF is also used to reimburse Mission Assignments (discussed in the next section) to other federal agencies. Current budgetary practice generally consists of funding the DRF through regular appropriations acts and then providing additional funding through supplemental appropriations when the account is deemed to be at risk of running out of funding. When the DRF balance becomes low, FEMA generally initiates a spending freeze on long-term recovery projects (primarily PA and HMGP projects), to save DRF funds to meet immediate, life-threatening needs. This policy remains in effect until the DRF is replenished with a new appropriation.[41] The DRF balance is not a criterion used by FEMA to recommend a major disaster declaration, nor does the President use the available balance to determine whether to issue a major disaster declaration.

CLOSEOUT

The termination of Stafford Act recovery projects are not subject to strict deadlines.[42] The requirements to "close out" a major disaster are established in 44 C.F.R. 13.50. In general, PA and HMGP projects take the longest to complete. Typically, major infrastructure projects take years to complete. Both the PA and HMPG have a common closeout sequence:

- Individual projects are managed by applicants and each project is separately closed by FEMA and the state when all of the costs associated with the project have been reconciled with supporting documentation.
- Disasters and emergencies are closed when all of the applicant's projects are closed and the applicant's administrative allowance expenses have been reconciled to supporting documentation.
- When all of the disaster's applicants are closed, the program can be closed. Programs are closed when the state's administrative costs have been reconciled with supporting documentation.
- When all programs within a disaster are closed, the disaster can be closed.[43]

Funding for recovery projects and activities carried out by federal agencies through Mission Assignments that are completed under budget is returned to the DRF. The state can make an appeal if they believe the project requires additional federal funding to be completed.[44]

Other Types of Federal Declarations

In some cases a major disaster declaration issued under the Stafford Act will trigger certain federal programs outside of FEMA. For instance, in certain circumstances a major disaster declaration will also trigger the Small Business Administration (SBA) disaster loan program. However, several federal departments and agencies have authorities to respond and declare certain types of disasters and emergencies separate from the Stafford Act. As with the above SBA example, these authorities may be exercised concurrently, or independently of FEMA. They may also be part of a response coordinated by the Secretary of Homeland Security under Homeland Presidential Directive-5. As mentioned previously, these authorities and responsibilities are carried out within the construct set forth in the National Response Framework (NRF), or under supplementary (known as "Annexes"), or complementary operational plans. Table 1 provides brief examples of other federal agencies and departments with separate authorities to respond to and declare certain types of disasters.

Table 1. Other Types of Federal Declarations

Scenario	Department/Agency	Authorities
Agriculture and Food	U.S. Department of Agriculture	The Secretary of Agriculture has the authority to declare an extraordinary emergency and take action due to the presence of a pest or disease of livestock that threatens livestock in the United States (7 U.S. Code § 8306 [2007]). The Secretary of Agriculture also has the authority to declare an extraordinary emergency and take action due to the presence of a plant pest or noxious weed whose presence threatens plants or plant products of the United States (7 U.S. Code § 7715 [2007]).
Public Health Emergency	Department of Health and Human Services	The Secretary of the Department of Health and Human Services has the authority to take actions to protect the public health and welfare, declare a public health emergency and to prepare for and respond to public health emergencies. (Public Health Service Act, 42 U.S.C. § 201, et seq. [2007]).
Oil and Hazardous Materials Spills	Environmental Protection Agency in the inland zone, or U.S. Coast Guard in the coastal zone	The EPA Administrator or Commandant of the USCG may classify an oil discharge as a as a Spill of National Significance. This designation allows for the appointment of a National Incident Commander. However, such designation does not provide additional funding or authority, nor does it alter response procedures. (40 CFR Part 300 [2006]).
Home and Business Disaster Loans	Small Business Administration	The Administrator of the Small Business Administration has the authority to provide home and business, and economic injury disaster loans in response to a gubernatorial request, the determination of a natural disaster by the Secretary of Agriculture, or based on the determination of the Secretary of Commerce that a fishery resource disaster or commercial fishery failure has occurred (P.L. 85-536, Section 7(b) 72 Stat. 387, as amended).

Source: Derived in part from exhibit 3 of Federal Emergency Management Agency/U.S. Department of Homeland Security, *Presidential Policy Directive/PPD-8*, Working Draft—National Response Framework, p. 17, at http://www.fema.gov/library/viewRecord.do?fromSearch=fromsearch&id=5450.

FEDERAL TO FEDERAL SUPPORT

Federal agencies and departments with responsibilities for handling incidents may work in conjunction with other federal entities when the scope of the work is beyond their normal operations. Federal to federal support under the Stafford Act is generally carried out through "Mission Assignments" while non-Stafford incidents, such as the ones described in Table 1, are carried out through other mechanisms such as interagency agreements and mutual aid agreements.

Stafford Act Incidents and Mission Assignments

Even when an emergency or major disaster is declared, FEMA does not exclusively perform all disaster response and recovery operations for the federal government. The President has the authority to direct any federal agency to use its authorities and resources in support of state and local response and recovery efforts under sections 402, 403, and 502 of the Stafford Act.

In general, when an emergency or major disaster declaration is declared FEMA coordinates federal entities and organizations that are involved in the incident by "assigning" missions to relevant agencies to address a state's request for federal assistance or support overall federal operations pursuant to, or in anticipation of a Stafford Act declaration. The activities carried out by other agencies through Mission Assignments are generally reimbursed by FEMA through the DRF.[45]

For example, FEMA may request the Department of Health and Human Services to establish and operate a shelter collocated with a federal medical station to support non-medical care givers and family members accompanying patients being treated at the station.

Non-Stafford Act Incidents

Federal agencies and departments that respond to incidents under the NRF may request and provide federal to federal support by executing interagency, intra-agency, or mutual aid agreements in accordance with applicable authorities.

OTHER KEY FEDERAL LAWS AND POLICIES

A number of statutory laws and policies supplement the Stafford Act. These statutes and policies organize and define the federal role in emergency preparedness, mitigation, response, and disaster relief. The following sections highlight some of these key laws and policies.[46]

Disaster Mitigation Act of 2000

P.L. 106-390, the Disaster Mitigation Act of 2000, amended the Stafford Act to authorize the President to establish a program of technical and financial assistance to states and local governments to assist in the implementation of pre-disaster hazard mitigation measures that are cost-effective and are designed to reduce injuries, loss of life, and property damage and destruction, including damage to critical services and facilities under the jurisdiction of the states or local governments.

The Disaster Mitigation Act of 2000 also established the National Predisaster Mitigation Fund and authorized the President to provide technical and financial assistance to each state and local government that has identified all natural disaster hazards in its jurisdiction and has demonstrated its ability to form effective public-private disaster hazard mitigation partnerships.

Post Katrina Emergency Management Reform Act

A wide range of reforms in federal emergency management were made in response to the 2005 hurricane season. Most of those reforms were included in Title VI of the DHS appropriations legislation for FY2007, the Post Katrina Emergency Management and Reform Act of 2006 (hereinafter the Post Katrina Act).[47]

The Post Katrina Act established new leadership positions and position requirements within FEMA, brought new missions into FEMA and restored some that had previously been removed. The Post Katrina Act also enhanced the agency's authority by directing the FEMA Administrator to undertake a broad range of activities before and after disasters occur. The Post Katrina Act contains provisions that set out new law, and amended the Stafford Act and the Homeland Security Act (hereinafter HSA).[48]

Among the provisions set forth by the Post Katrina Act were certain preparedness requirements. For example, the President was authorized to establish a national preparedness goal and national preparedness system, and to complete, revise, and update (as necessary) the goal to ensure the nation's ability to prevent, respond to, recover from, and mitigate against disasters of all kinds, including acts of terrorism.[49] The goal must be consistent with NIMS and the National Response Plan (now the NRF).[50]

The Post Katrina Act also established reporting requirements intended to ensure that Congress receives information on the implementation of specified new policies, including those directed at ensuring that a qualified workforce exists, and guidelines for enhanced aid to individuals and families are developed.[51] Other provisions included oversight and accountability measures to audit expenditures and develop and maintain internal management controls to detect and eliminate fraud, waste, and abuse.

Homeland Security Presidential Directives

In response to the September 11th terrorist attacks, President Bush signed a number of Homeland Security Presidential Directives (HSPD) that have influenced federal emergency management policies.[52] Key among these are HSPD-8 (see "The Framework Approach") signed on December 17, 2003, and HSPD-5 signed on February 28, 2003. The purpose of HSPD-5 is to enhance the ability of the United States to manage domestic incidents by establishing a single, comprehensive national incident management system. HSPD-5 mandated the DHS Secretary to develop a national response plan, and a concept of operations that would incorporate all levels of government within a single unifying framework to manage domestic incidents through the implementation of NIMS. Under HSPD-5, all federal agencies are required to adopt NIMS.

Homeland Security Act

On November 25, 2002, President Bush signed into law the Homeland Security Act of 2002 (hereinafter HSA).[53] HSA authorized the reorganization of various federal agencies and established the Department of Homeland Security (DHS). In addition to the creation of DHS, HSA made several changes to existing federal programs and created new ones. Most of these changes were related to terrorism. However, others had an influence on emergency management. Section 501 of Title V established in DHS the Directorate of Directorate of Emergency Preparedness and Response. Section 502 makes the DHS Secretary responsible for helping to ensure the effectiveness of emergency response providers to terrorist attacks, major disasters, and other emergencies, aiding recovery from terrorist attacks and major disasters, consolidating existing federal government emergency response plans into a single, coordinated national response plan, and developing comprehensive programs for developing interoperative communications technology and helping to ensure that emergency response providers acquire such technology.

National Oil and Hazardous Substances Pollution Contingency Plan[54]

The National Oil and Hazardous Substances Pollution Contingency Plan, commonly referred to as the National Contingency Plan (NCP) for short, is the federal government's principal plan for responding to oil spills and releases of hazardous substances.[55] It provides the framework for coordinating the federal, state, and local roles in responding to such incidents and notifying the relevant agencies. The procedures of the NCP are codified in federal regulation and are binding and enforceable.[56] In contrast, most federal emergency response plans are only administrative frameworks internal to the federal government for guiding agency roles and responsibilities, making the NCP somewhat unique in this respect among federal emergency response plans.

Three federal laws authorized the development of the regulations that are embodied in the NCP. The first two, the Oil Pollution Act (OPA) and Section 311 of the Clean Water Act, authorize federal emergency response to oil spills into U.S. waters, onto adjoining shorelines, or that may affect natural resources under the jurisdiction of the United States.[57] The third law, the Comprehensive, Environmental Response, Compensation, and Liability Act (CERCLA, commonly referred to as Superfund) authorizes federal emergency response to releases of hazardous substances into the environment.[58] The term "environment" includes surface and subsurface lands, surface waters, groundwater, and ambient air, making the response authorities for hazardous substances broader in terms of their physical reach than that for oil spills. Other response authorities apply to oil released under certain circumstances not covered by the NCP.[59]

The President's response authorities under these three laws are delegated by executive order to the Environmental Protection Agency (EPA) in the inland zone and to the U.S. Coast Guard in the coastal zone.[60] The inland zone is the environment inland of the coastal zone, excluding the Great Lakes and specified ports and harbors on inland rivers which are part of the coastal zone.[61] The above laws also directed the President to develop the regulations of the NCP to establish procedures for carrying out response actions. Consistent with the executive orders, the NCP designates EPA and the U.S. Coast Guard as the lead federal agencies within their respective zones, and outlines the roles of other supporting federal agencies and state and local governments that may participate in a federal response. State and local officials often act as the "first responders" under the NCP to initiate measures to protect public safety.[62]

Two dedicated trust funds finance the costs of federal response actions carried out under the NCP.[63] The U.S. Coast Guard administers the Oil Spill Liability Trust Fund to finance the costs of responding to oil spills. EPA administers the Hazardous Substance Superfund Trust Fund to finance the costs of responding to releases of hazardous substances. The federal government may recover its response costs from the responsible parties, which are to be deposited back into the respective trust fund to finance future response actions. The responsible parties also may use their own funds to carry out response actions under federal oversight. If the responsible parties cannot be found or cannot pay, federal response costs ultimately are borne by the taxpayer.[64]

Once an emergency response is complete, the NCP also governs any long-term remediation of environmental contamination that may be needed. Long-term remediation may continue for several years, even decades in some cases, depending on the nature and extent of the contamination. For example, the remediation of contamination in groundwater typically is a lengthy and technically complex process lasting multiple years. In some cases, the federal response also may entail the long-term restoration of natural resources affected by an oil spill or a release of a hazardous substance.

Although the NCP is most often used as a stand-alone authority, the Secretary of Homeland Security may invoke the NCP under the National Response Framework in situations that may warrant coordination with other federal response plans. Such situations may include a major disaster or emergency declared under the Stafford Act, and other multi-faceted incidents. ESF#10 of the National Response Framework—Oil and Hazardous Materials Annex—applies the NCP as the operative federal response plan for incidents involving an oil spill or a release of "hazardous materials" (i.e., hazardous substances, as termed in the NCP). Regardless of whether the NCP is applied as a stand-alone authority or through the structure of the National Response Framework, the procedures for responding to an oil spill or a release of a hazardous substance are the same because the NCP remains the operative federal response plan in either instance.

KEY FEDERAL ASSISTANCE FOR DISASTER RESPONSE AND RECOVERY

As mentioned earlier in this report, while the majority of emergency and disaster assistance is coordinated and provided by FEMA, other federal agencies provide assistance under authorities in conjunction with the Stafford Act declaration or through separate authorities. Several of these agencies are able to provide assistance without a presidential declaration. The concluding section highlights of some of the key agencies that provide emergency and disaster assistance to states and localities including a brief description of the work performed by each agency.

National Guard[65]

The National Guard is both a state and federal organization: it is simultaneously the organized militia of a state or territory and a reserve component of the Army and the Air Force. Due to its size (over 450,000 individuals), trained personnel, and available equipment, it is frequently used for emergency response. Normally, the National Guard operates in a state status, under the control of state and territorial governors, who can order National Guard personnel to perform full-time "state active duty" in response to disasters and civil disorders. In this state capacity, National Guard personnel are not subject to the restrictions of the Posse Comitatus Act (that is, they can perform law enforcement functions).

The National Guard can also be activated under the authority of 32 U.S.C. 502(f). This provision of federal law provides that "a member of the National Guard may ... without his consent, but with the pay and allowances provided by law ... be ordered to perform training or other duty in addition to [inactive duty for training or annual training]." The advantage of using this authority is that the National Guard personnel called will receive federal pay and benefits and are entitled to certain legal protections as though they were in federal service, but they remain under the control of their state or territorial governor and are therefore not subject to the restrictions of the Posse Comitatus Act. Such activation can only occur with the approval of the Department of Defense.[66]

Finally, the National Guard can be activated in a purely federal status. When federally activated, National Guard personnel operate under the control of the President, receive federal pay and benefits and are subject to the Posse Comitatus Act.[67]

Department of Agriculture[68]

The U.S. Department of Agriculture (USDA) has several agencies that provide emergency and disaster assistance, including the Farm Service Agency, the Natural Resources Conservation Service, and Rural Development. The majority of USDA assistance does not require a Stafford declaration. Rather, depending on the program, decisions regarding assistance are typically made by the Secretary (e.g., emergency loans). Assistance may also be provided to participating producers under standing programs (e.g., federally subsidized crop insurance).

Assistance for agricultural producers includes emergency loans, direct payments under the noninsured crop disaster program, and crop insurance.[69] Disaster assistance is also provided for impairments to watersheds, including debris removal from stream channels, road culverts, and bridges, repairing levees, reseeding damaged areas, and purchasing floodplain easements. Private land assistance includes repairs to damaged farmland and re-establishment of forest land.[70] Community assistance is also available in the form of housing disaster loans and grants, moratorium on loan payments, loan servicing assistance for essential community facilities, and emergency community water assistance grants.[71]

Department of Defense[72]

The role of the Department of Defense (DoD) and its components in supporting domestic response efforts is potentially significant. DoD has a broad range of capabilities that could be useful to civil authorities in emergency situations, including transportation assets, medical personnel and supplies, security forces, and communications equipment. However, the extent of the support DoD can provide is limited by the need to conduct its principal mission of national defense. The two principal ways in which DoD can provide defense support of civil authorities (DSCA) are by way of an "immediate response," or in response to a formal "request for assistance" (RFA). Additionally, as part of a proactive federal response, DoD could preposition assets in anticipation of a request.

Immediate Response

Certain defense officials can provide DSCA in a limited manner using "immediate response" authority. Immediate response authority enables local military commanders and certain DoD civilians to act immediately to "save lives, prevent human suffering, or mitigate great property damage within the United States" when they receive a request for assistance from a civil authority under conditions that are "imminently serious" and when time "does not permit approval from higher authority."[73] Assistance provided under this authority might include providing medical care, restoring critical public services, and distributing food and other supplies. However, commanders using this authority are obligated to seek approval or authorization through their chain of command as soon as possible.

Requests for Assistance

The Department of Defense may also provide support in response to a "request for assistance" from state, local, or tribal entities, or from another federal agency.[74] The Secretary of Defense has the principal authority for approving DSCA requests and retains control of all DoD assets provided. The forces provided in response to an RFA could include units from multiple military services and could include activated members of the National Guard and Reserve. Members of the federal reserves[75] are always activated after a request for assistance under Title 10 of the U.S. Code. Therefore when acting in a federal capacity, the National Guard operates under the control of the President, receives federal pay and benefits, and is subject to the Posse Comitatus Act in the same manner as active duty military personnel. National Guard forces remain under the control of their respective governors unless ordered into federal service (see the previous section on the National Guard). There are a number of statutory authorities which can be used to activate members of the National Guard and Reserve, although they all have constraints that could potentially limit their utility in providing reserve forces for domestic response missions.[76]

Army Corps of Engineers[77]

Under its Civil Works Program, the U.S. Army Corps of Engineers (Corps) performs emergency response actions to respond to flood and coastal emergencies, droughts, and other disasters.[78] The Corps also plans, builds, operates, and maintains a wide range of water resources facilities, including hurricane protection and flood damage reduction projects. The Stafford Act authorizes FEMA to direct the Department of Defense (DoD), which generally works through the Corps, to provide assistance in the event of an emergency or major disaster. The Corps provides assistance to monitor and stabilize damaged structures and demolish structures designated as immediate hazards to public health and safety. Other forms of assistance include technical assistance in clearing, removing, and disposing of contaminated and uncontaminated debris from public property, and establishing ground and water routes into affected areas.[79] The Corps also has authorities specific to the type of natural disaster that has occurred. After a flooding event, the Corps has the authority to rehabilitate certified flood control works (e.g., levees) and federally constructed hurricane or shore protection projects and to conduct related inspections.[80] In the case of drought emergencies, the Corps may allow for temporary water withdrawal from reservoirs for municipal and industrial use, as well as the construction of wells for transport of water to drought-distressed areas.[81]

Department of Health and Human Services[82]

The Department of Health and Human Services (HHS) provides grants to states and localities to strengthen public health and medical system capacity in preparedness for health threats.[83] Under the NRF, the department serves as the primary agency for ESF #8 (Public Health and Medical Services) to respond to the public health and medical consequences resulting from natural disasters and human-caused incidents. The Secretary of HHS has general authority under the Public Health Service Act to provide public health assistance to states, upon their request, without a Stafford Act declaration. In addition, the Secretary has authority to determine that a public health emergency exists.[84]

Pursuant to such a determination, the Secretary may waive certain administrative requirements, provide additional forms of assistance, and take certain other actions to expand federal aid to state and local governments, not-for-profit entities, and others. In either case, however, the Secretary lacks a dedicated funding mechanism for the provision of assistance.

Department of Housing of Housing and Urban Development[85]

The Department of Housing and Urban Development (HUD) may play a variety of roles in disaster recovery. When the President declares a major disaster, HUD may offer waivers of various program rules and regulations to allow communities to use their existing HUD program funding and resources to meet needs arising from the disaster.[86] In some years, Congress has provided additional appropriations for these programs—particularly the Community Development Block Grant (CDBG) program—to be used specifically for disaster assistance and recovery. Similarly, in some years, Congress has provided HUD with funding to provide housing vouchers to disaster displaced survivors, in conjunction with FEMA, or directed FEMA to contract with HUD for the provision of such vouchers. HUD, through the Federal Housing Administration (FHA), may also offer special mortgage insurance products to and special forbearance for borrowers affected by major disasters.[87]

Community Development Block Grants

HUD's Community Development Block Grant (CDBG) program provides formula grants to eligible communities to use for a range of housing and community development activities. The CDBG program has been used frequently by the federal government to respond to natural and human-caused incidents. In general, Congress has provided increased flexibility and allocated additional CDBG funds to affected communities and states to help them mitigate against, respond to, and recover from presidentially declared disasters.[88]

Department of Transportation[89]

The U.S. Department of Transportation (DOT) works with FEMA to coordinate transportation support to disaster stricken areas.[90] DOT reports on damage to transportation infrastructure, coordinates alternative transportation services, and coordinates the restoration and recovery of the transportation infrastructure, as well as other emergency responses involving transportation.

DOT's Federal Highway Administration's (FHWA) Emergency Relief Program (ER) is the Department's disaster program for highways.[91] The ER program is administered through the state departments of transportation in close coordination with FHWA's division offices.[92] While ER is a federal program, the decision to seek ER funding is made by the state rather than the federal government.

The ER program provides funds for the repair and reconstruction of roads on the federal-aid highway system that have suffered serious damage as a result of either (1) a natural disaster, such as a flood, hurricane, tidal wave, earthquake, tornado, severe storm, or landslide; or (2) a catastrophic failure from any external cause.[93]

In 2012, Congress passed the Moving Ahead for Progress in the 21st Century Act (MAP-21; P.L. 112-141). Section 20017 of the Act included a new disaster relief program for mass transit, the Public Transportation Emergency Relief Program.

Once established, the program is to fund "operating costs" for evacuation services; rescue operations; temporary public transportation service; or reestablishing, expanding, or relocating public transportation route service before, during, or after an emergency. The program is to be funded from the General Fund on a such sums as necessary basis and will require appropriations action. Most observers expect the new program will be administered by the Federal Transit Administration.

Federal Highway Administration

As noted above, the FHWA has a permanent program dedicated to disaster relief, the ER program. Historically, the vast majority of ER funds have gone for natural disaster repair and reconstruction. For example, since 2005, ER funding allocations for Gulf Coast Hurricane response have totaled over $3 billion.

Funding provided for hurricane relief includes funds from the Program's annual $100 million highway trust fund authorization and from additional appropriations provided in supplemental or other appropriations acts. Generally, ER funds can only be used for roads and bridges on the federal-aid highway system. Repairs and reconstruction costs for other damaged roads (mostly local roads and neighborhood streets) may be reimbursed by FEMA.

Environmental Protection Agency[94]

The U.S. Environmental Protection Agency (EPA) is most frequently involved in emergency response as the lead federal agency responsible for responding to oil spills and releases of hazardous substances within the inland zone under the National Contingency Plan. EPA also serves in this capacity if this plan is invoked under the National Response Framework through ESF #10—Oil and Hazardous Materials Annex. (See the section on the "National Contingency Plan" in this report.) EPA also has various other response roles under the National Response Framework and may perform a wide array of support functions in responding to a disaster or emergency.[95] EPA support to other federal agencies and state and local governments broadly includes activities necessary to address threats to human health and the environment focusing on impacts to drinking water and wastewater treatment facilities and post-disaster cleanup.

In accordance with various ESFs under the NRF, as well as Homeland Security Presidential Directive 7: Critical Infrastructure Identification, Prioritization, and Protection (HSPD-7), EPA may provide a wide range of support functions to other federal agencies (primarily FEMA and the Army Corps of Engineers) and state and local governments to protect drinking water or wastewater facilities.[96] As needed, EPA may provide guidance and technical assistance to determine impacts from firefighting chemicals on wastewater treatment facilities; safely manage contaminated debris; identify critical water and wastewater systems requiring priority power restoration; or identify alternate water supplies and wastewater collection and treatment for critical health care facilities.

EPA also has additional emergency response roles related to protecting water infrastructure under other response plans and authorities. EPA is the lead federal agency for the water sector under the National Infrastructure Protection Plan.[97] EPA also has statutory "emergency powers" under the Safe Drinking Water Act to issue orders and commence civil action if a contaminant likely to enter a public water supply system poses a substantial threat to public health, and state or local officials have not taken adequate action.[98] The Public Health Security and Bioterrorism Preparedness and Response Act of 2002 amended these emergency powers to include the authority for EPA to take actions in response to a threatened or potential terrorist attack, or other intentional act, that would disrupt the provision of safe drinking water or impact the safety of a community's water supply.[99]

EPA may play a prominent role in post-disaster cleanup under the NRF.[100] EPA may support the Army Corps of Engineers in its mission under ESF #3—Public Works and Engineering Annex— to remove disaster debris.[101] The removal of contaminated debris, and hazardous materials comingled with debris, is coordinated under ESF #10—Oil and Hazardous Materials Annex. In responding to such incidents, EPA may provide technical assistance to state solid waste agencies regarding proper hazardous and nonhazardous waste management, including identifying recycling/reuse opportunities to minimize disaster debris and identifying potential waste staging or storage areas and disposal facilities. EPA also may support the U.S. Department of Energy's responsibility to maintain continuous and reliable energy supplies for the United States during disaster recovery under the National Response Framework through ESF #12—Energy Annex. In practice, EPA support for this function has involved waiving environmental requirements applicable to motor vehicle fuel under the Clean Air Act. As part of the federal response to hurricanes in 2005, EPA granted certain waivers under this statute in response to requests from state and local officials when significant disruptions in fuel production or distribution occurred in the wake of these natural disasters.[102]

Forest Service[103]

The Forest Service (FS) is one of two federal entities responsible for protecting most federal lands from wildfire—a situation that often requires immediate assistance.[104] Wildfire management is generally split into three categories: preparedness, suppression, and fuel reduction. The FS has received more than $2.0 billion annually over the last few years for wildfire management.[105] Federal wildfire policy is to actively suppress all wildfires and to protect lives, property, and resources on federal lands.[106]

In an emergency wildfire situation, the FS may work with multiple agencies. The National Interagency Fire Center (NIFC)—of which the FS is a member of—coordinates federal, state, and private forces to assist the state or region in need for complex situations while maintaining local wildfire protection.[107] As mentioned previously, there are FMAGs available from FEMA that provide grants, equipment, personnel, and supplies to supplement community resources when fires threaten destruction that might warrant a major disaster declaration. If the President declares a disaster, fire management assistance and other recovery programs are also available from FEMA under the Stafford Act.[108] The type of assistance depends on a number of factors, such as the nature and severity of the wildfire damages and the insurance coverage of the affected parties. Also, the FS is involved in site rehabilitation and restoration following a wildfire via the Burned Area Emergency Response (BAER) program. Under BAER, emergency treatments on federal lands are prescribed and implemented to minimize threats to life or property resulting from the effects of a fire or to stabilize and prevent unacceptable degradation to natural and cultural resources.

Small Business Administration

Since its creation in 1953, the Small Business Administration (SBA) Disaster Loan Program has offered direct loans to individuals and businesses to help repair, rebuild, and recover from economic losses after a disaster. The majority of disaster loans (approximately 80%) are made to individuals and households (renters and property owners) to help repair and replace homes and personal property. The three main types of loans for disaster-related losses provided by the SBA Disaster Loan Program are (1) Home Disaster Loans, (2) Business Disaster Loans, and (3) Economic Injury Disaster Loans (EIDL).

Home disaster loans are used by homeowners and renters to repair or replace their disaster-damaged primary residences or personal property. SBA regulations limit home loans to $200,000 for the repair or replacement of real estate and $40,000 for the repair or replacement of personal property. Business disaster loans help businesses of all sizes and nonprofit organizations repair or replace disaster-damaged property, including inventory and supplies. EIDLs provide working capital to businesses that have been impacted by an incident. EIDLs cannot be used to restore or repair businesses. Both Business Disaster Loans and EIDLs are limited by law to $2 million per applicant. They also provide assistance to small agricultural cooperatives (but not enterprises), and certain private, nonprofit organizations that have suffered substantial economic injury resulting from a physical disaster or an agricultural production disaster.[109]

End Notes

[1] William L. Waugh, Jr., Living with Hazards, Dealing with Disasters An Introduction to Emergency Management (Armonk, NY: M.E. Sharpe, 2000).

[2] Homeland Security Presidential Directive 5: Management of Domestic Incidents, http://www.dhs.gov/xabout/laws/ gc_1214592333605.shtm

[3] For example, prior to ICS police and fire departments responding to the same incident might use different radio frequencies and communicate with different terms. A "code blue" for one department might mean something else for another. The organization structure might also be different. A commander in one department might have a different role and responsibility in another. ICS (and NIMS) is therefore an attempt to eliminate potential confusion caused by these differences.

[4] A glossary of emergency and disaster related terms can be located at http://www.fema.gov/ glossary.

[5] For further analysis on PPD-8 see CRS Report R42073, Presidential Policy Directive 8 and the National Preparedness System: Background and Issues for Congress, by Jared T. Brown.

[6] HSPD-8 gave the Secretary of the Department of Homeland Security broad authority to improve preparedness, prevention, response, and recovery operations among others. HSPD-8 also required the creation of a National Preparedness Goal and the implementation of a National Preparedness System. HSPD-8 fulfilled many of the requirements later set forth in the Post Katrina Emergency Management Reform Act of 2006 (Title VI of the Department of Homeland Security Appropriations Act, 2007—hereinafter the Post-Katrina Act) such as actions to approach all hazards within a risk-based framework, and the use of metrics to measure levels of preparedness. Many such requirements, however, were not implemented.

[7] The White House, Presidential Policy Directive/PPD-8, March 30, 2011, p. 1, at http://www.dhs.gov/xlibrary/assets/ presidential-policy-directive-8-national-preparedness. pdf.

[8] P.L. 109-295, 6 U.S.C. §741- 764, 120 STAT. 1424-1433.

[9] The White House, Presidential Policy Directive/PPD-8, March 30, 2011, p. 3, athttp://www.dhs.gov/xlibrary/assets/ presidential-policy-directive-8-national-preparedness. pdf.

[10] Ibid, p. 3.

[11] These descriptions should not be viewed as definitive. Some of the planning frameworks are being updated or still in draft form. They could conceivably change in form or content before being implemented.

[12] Federal Emergency Management Agency, Presidential Policy Directive 8: National Preparedness: Working Drafts of the National Planning Frameworks - Prevention Framework, p. 9.

[13] Ibid., p. 10.

[14] Ibid., p. 1.

[15] Ibid, p. 12.

[16] For further information on the NRF see CRS Report RL34758, The National Response Framework: Overview and Possible Issues for Congress, by Bruce R. Lindsay.

[17] U.S. Federal Emergency Management Agency, ESF Annexes Introduction, January 2008, at http://www.fema.gov/ pdf/emergency/nrf/nrf-esf-intro.pdf.

[18] U.S. Federal Emergency Management Agency, Support Annexes: Introduction, January 2008, at http://www.fema.gov/pdf/emergency/nrf/nrf-support-intro.pdf.

[19] U.S. Federal Emergency Management Agency, Incident Annexes Introduction: National Response Plan, December 2004, pp. INC-i, at http://www.learningservices.us /pdf/emergency/nrf/nrp_incidentannexintroduction.pdf.

[20] Federal Emergency Management Agency, National Disaster Recovery Framework: Strengthening Disaster Recovery for the Nation, September 2011, p. 1,at http://www.fema.gov/pdf/recoveryframework/ndrf.pdf.

[21] Codified at 42 U.S.C. §5121 et seq. For further analysis on the Stafford Act see CRS Report RL33053, Federal Stafford Act Disaster Assistance: Presidential Declarations, Eligible Activities, and Funding, by Francis X. McCarthy.

[22] P.L. 93-288, 42 U.S.C. §5191(b). The President is required to consult with the state governor when practical. Examples of these declarations include the April 19, 1995, bombing of the Alfred P. Murrah Building in Oklahoma City, and the September 11, 2001, attack on the Pentagon in Virginia.

[23] For further analysis on emergency and disaster declarations see CRS Report RL34146, FEMA's Disaster Declaration Process: A Primer, by Francis X. McCarthy, and CRS Report R41981, Congressional Primer on Major Disasters and Emergencies, by Francis X. McCarthy and Jared T. Brown.

[24] There are ten FEMA regions in the United States (including the U.S. territories). The authority to issue a FMAG through the FEMA Regional Director is designed to make the issuance of a FMAG as quickly as possible.

[25] P.L. 93-288, 42 U.S.C. §5122(2).

[26] 44 C.F.R. §206.48.

[27] For more detail see the complete Public Assistance Policy Reference Manual at http://www.fema.gov/pdf/ government/grant/pa/policy.pdf. Private businesses are not eligible for PA.

[28] The maximum amount is been modified to reflect inflationary change.

[29] Title VI of P.L. 109-295, the FY2007 DHS appropriations legislation.

[30] P.L. 109-295, §684, Stafford Act amended §404, 120 Stat. 1447.

[31] Emergency declarations are different from a "state of emergency." A state of emergency is generally issued by the state governor and varies from state to state. States of emergency usually trigger the activation or notification of certain entities at the state level to respond to a threat or incident.

[32] P.L. 93-288, 42 U.S.C. §5122(1).

[33] Recent examples of pre-event declarations include emergency declarations prior to Hurricanes Katrina, Rita, and Gustav making landfall (emergency declarations 3212, 3260, and 3290 respectively).

[34] P.L. 93-288, 42 U.S.C. §5187(a).

[35] P.L. 93-288, 42 U.S.C. §5170b.

[36] 42 U.S.C. 5172(b).

[37] Per 44 CFR §206.47, a state must have accumulated more than $125 in damages per capita within the state to reach a traditional cost-share waiver. The cost-share cannot be waived administratively for all federal Stafford Act disaster assistance programs. For example, the Hazard Mitigation Grant Program and the Other Needs Assistance program would require a statutory waiver of the existing statutory cost-share.

[38] For further analysis on cost-shares see CRS Report R41101, FEMA Disaster Cost-Shares: Evolution and Analysis, by Francis X. McCarthy.

[39] For further analysis on the DRF see CRS Report R40708, Disaster Relief Funding and Emergency Supplemental Appropriations, by Bruce R. Lindsay and Justin Murray.

[40] While most appropriations expire after a set period of time, no-year appropriations are available until expended. This is helpful in disaster recovery since infrastructure repair and mitigation projects can stretch out over several years.

[41] P.L. 112-25, the Budget Control Act of 2011 (hereafter the BCA) provides a mechanism designed, arguably, to limit spending on major disasters declared under the Stafford Act. For further analysis of disaster relief under the BCA see CRS Report R42352, An

Examination of Federal Disaster Relief Under the Budget Control Act, by Bruce R. Lindsay, William L. Painter, and Francis X. McCarthy.

[42] Applications for assistance are subject to deadlines. While some assistance has suggested end dates, the President has discretion to extend the deadline (e.g., temporary housing).

[43] Richard L. Skinner, Opportunities to Improve FEMA's Disaster Closeout Process, U.S. Department of Homeland Security Office of Inspector General, OIG-10-49, Washington, DC, January 27, 2010, p. 4, at http://www.oig.dhs.gov/ assets/Mgmt/OIG_10-49_Feb10.pdf.

[44] See Federal Emergency Management Agency, Program Management and Grant Closeout: Standard Operating Procedure, p. 30, at http://www.fema.gov/pdf/government /grant/pa /9570_14.pdf.

[45] Funds from the DRF are not available for activities undertaken under other authorities or agency missions, or for non-Stafford Act incidents requiring a coordinated federal response.

[46] The Public Health Service Act is discussed in the Department of Health and Human Services section of this report. See also Table 1. The Small Business Act is discussed in Table 1.

[47] P.L. 109-295.

[48] P.L. 107-296, 116 Stat. 2140-2321, as amended, 6 U.S.C. 101 et seq.

[49] P.L. 109-295, §§ 642 and 643, 120 STAT. 1425.

[50] P.L. 109-295, § 643(b) 120 STAT. 1425.

[51] P.L. 109-295, §§ 642 and 643, 120 STAT. 1425.

[52] See U.S. Congress, House Committee on Homeland Security, Compilation of Homeland Security Directives (HSPD), Updated through December 31, 2007), 110th Cong., 2nd sess., December 31, 2007, Committee Print110–B (Washington: GPO, 2008).

[53] P.L. 107-296.

[54] This section was authored by David M. Bearden, Specialist in Environmental Policy, Resources, Science, and Industry Division.

[55] The NCP also applies to releases of pollutant or contaminants that may present an imminent and substantial danger to public health or welfare.

[56] 40 C.F.R. Part 300.

[57] 33 U.S.C. § 2701 et. seq, and 33 U.S.C. § 1321, respectively. For further discussion of the authorities of OPA and Section 311 of the Clean Water Act, see CRS Report RL33705, Oil Spills in U.S. Coastal Waters: Background and Governance, by Jonathan L. Ramseur.

[58] 42 U.S.C. § 9601 et. seq. For further discussion of the authorities of CERCLA, see CRS Report R41039, Comprehensive Environmental Response, Compensation, and Liability Act: A Summary of Superfund Cleanup Authorities and Related Provisions of the Act, by David M. Bearden.

[59] Subtitle I of the Solid Waste Disposal Act addresses petroleum leaked from underground storage tanks. This role is performed mainly by the states under cooperative agreements with EPA.

[60] Executive Order 12580 delegated the President's authorities under CERCLA, and Executive Order 12777 delegated the President's authorities under OPA and Section 311 of the Clean Water Act. Executive Order 13286 amended these executive orders to reflect the transfer of the U.S. Coast Guard from the Department of Transportation to the Department of Homeland Security in 2003.

[61] 40 C.F.R. § 300.5.

[62] 40 C.F.R. § 300.180(f).

[63] Response costs for incidents occurring on federal facilities or vessels are funded with separate appropriations to the agencies with administrative jurisdiction over those facilities or vessels.

[64] Currently, the vast majority of the revenues for the Oil Spill Liability Trust Fund are derived from a dedicated eight cents per-barrel tax on domestic and imported oil. The tax is scheduled to terminate at the end of 2017. The Hazardous Substance Superfund Trust Fund is financed mostly with revenues transferred from the General Fund of the U.S. Treasury, since the taxes on domestic and imported oil, chemical feedstocks, and corporate income that once financed this trust fund expired at the end of 1995.

[65] This section was authored by Lawrence Kapp, Specialist in Military Manpower Policy, Foreign Affairs, Defense, and Trade Division.

[66] See the section on Department of Defense for further information on this topic.

[67] For further analysis on the use of Reserve Personnel see also CRS Report RL30802, Reserve Component Personnel Issues: Questions and Answers, by Lawrence Kapp.

[68] This section was authored by Dennis Shields, Specialist in Agricultural Policy; Megan Stubbs, Specialist in Agricultural Conservation and Natural Resources Policy; and Tadlock Cowan, Analyst in Natural Resources and Rural Development, Resources, Science and Industry Division.

[69] For more information, see CRS Report RS21212, Agricultural Disaster Assistance, by Dennis A. Shields. For a summary of USDA Farm Service agency disaster programs, see http://www.fsa.usda.gov/Internet/FSA_File/ emerfsaprograms.pdf. For information on crop insurance, see CRS Report R40532, Federal Crop Insurance: Background and Issues, by Dennis A. Shields.

[70] See CRS Report R40763, Agricultural Conservation: A Guide to Programs, by Megan Stubbs for information on watershed assistance (p. 14) and private land assistance (p.13). See CRS Report RL31065, Forestry Assistance Programs, by Megan Stubbs, for information on forestry programs (p. 17).

[71] For information on Rural Development programs, see http://www.rurdev.usda.gov/rd/disasters /disassistance.html and CRS Report RL31837, An Overview of USDA Rural Development Programs, by Tadlock Cowan (p.31).

[72] This section was authored by Lawrence Kapp, Specialist in Military Manpower Policy, Foreign Affairs, Defense, and Trade Division.

[73] Department of Defense, Department of Defense Directive 3025.18, "Defense Support of Civil Authorities," Washington, DC, December 29, 2010, para. 4.g., at http://www.dtic.mil /whs/directives/corres/pdf/302518p.pdf.

[74] For a summary of how the Department of Defense handles requests for assistance, see Joint Publication 3-28, Civil Support, September 14, 2007, pages II-3 to II-7, at http://www.dtic.mil/doctrine/new_pubs/jp3_28.pdf .

[75] The Army Reserve, Navy Reserve, Marine Corps Reserve, and Air Force Reserve. The Coast Guard Reserve normally operates under the control of the Secretary of Homeland Security.

[76] The principal statutes for activating members of the reserve component are 10 USC § 12301(a), 12301(b), 12301(d), 12302, 12304, 12304a and 12304b. Each of these authorities contains constraints, including whether Congress, the President, or the Service Secretary can invoke it; how many reservists can be activated; how long reservists can be activated; the purposes for which the reservists may be used; and/or limitations on the activation of National Guard forces. The authority most likely to be used in the event of a major disaster or emergency is the recently enacted 10 USC 12304a, which permits activation of members of the Army Reserve, Navy Reserve, Marine Corps Reserve and Air Force Reserve for up to 120 days when a governor requests federal assistance in responding to a major disaster or emergency. National Guard personnel can also be called into federal service under 10 USC § 331-333 and § 12406 for certain purposes, including the suppression of insurrection

against a state government, the enforcement of federal laws, and the prevention of interference with state and federal laws, if that interference deprives a class of people of rights, privileges, immunities, or protections named in the Constitution.

[77] This section was authored by Charles Stern, Specialist in Natural Resources Policy, Resources, Science and Industry Division.

[78] The Corps is an agency within the Department of Defense with military and civilian missions. For more information on Corps emergency authorities, see CRS Report R41243, Army Corps of Engineers Water Resource Projects: Authorization and Appropriations, by Nicole T. Carter and Charles V. Stern.

[79] Contaminated debris management is coordinated with the Environmental Protection Agency. See CRS Report RL33477, Disaster Debris Removal After Hurricane Katrina: Status and Associated Issues, by Linda Luther.

[80] This program is known as the Rehabilitation and Inspection Program. For more information, see CRS Report R41752, Locally Operated Levees: Issues and Federal Programs , by Natalie Keegan et al.

[81] 33 U.S.C. 708 and 33 U.S.C. 701n. In the case of reservoir withdrawals, Administration policy limits application to governor-declared drought emergencies and up to 99 acre-feet.

[82] This section authored by Sarah A. Lister, Specialist in Public Health and Epidemiology, Domestic Social Policy Division. For further analysis on public health and medical emergencies see CRS Report RL33579, The Public Health and Medical Response to Disasters: Federal Authority and Funding, by Sarah A. Lister.

[83] See HHS, Public Health and Emergency Preparedness cooperative agreements (a type of grant), at http://www.cdc.gov/phpr/coopagreement.htm; and Hospital Preparedness Program, at http://www.phe.gov/ preparedness/planning/hpp/pages/default.aspx.

[84] Public Health Service Act §319; 42 U.S.C. 247d.

[85] This section authored by Maggie McCarty, Specialist in Housing Policy, Domestic Social Policy Division.

[86] For example, communities may amend their community development plans and reprogram existing Community Development Block Grant funds without prior HUD approval, citing the "urgent need" objectives of program.

[87] For more information on the role of HUD in disaster response see CRS Report RS22358, The Role of HUD Housing Programs in Response to Hurricane Katrina, by Maggie McCarty, Libby Perl, and Eugene Boyd; CRS Report RL33078, The Role of HUD Housing Programs in Response to Past Disasters, by Maggie McCarty, Libby Perl, and Bruce E. Foote; and CRS Report RL33330, Community Development Block Grant Funds in Disaster Relief and Recovery, by Eugene Boyd.

[88] For more information about CDBG, see CRS Report RL33330, Community Development Block Grant Funds in Disaster Relief and Recovery, by Eugene Boyd.

[89] This section authored by Robert Kirk, Specialist in Transportation Policy, Resources, Science and Industry Division.

[90] U.S. Department of Homeland Security, National Response Framework. "Emergency Support Function #1 – Transportation Annex Washington, at http://www.fema.gov/pdf/emergency /nrf/nrf-esf-01.pdf .

[91] For more information about the ER program see CRS Report R42021, Emergency Relief Program: Federal-Aid Highway Assistance for Disaster-Damaged Roads and Bridges, by Robert S. Kirk, and CRS Report RS22268, Repairing and Reconstructing Disaster-Damaged Roads and Bridges: The Role of Federal-Aid Highway Assistance, by Robert S.

Kirk. Under certain circumstances the Federal Aviation Administration (FAA) and the Federal Transit Administration (FTA) provide disaster assistance.

[92] There is one in each state.

[93] For example, a bridge collapse.

[94] This section is authored by David M. Bearden, Specialist in Environmental Policy, and Linda Luther, Analyst in Environmental Policy, Resources, Science, and Industry Division.

[95] For more information about EPA responsibilities under the National Response Framework, including those under individual ESFs, see EPA's "Federal Response Plans" website at http://www.epa.gov/homelandsecurityportal/lawsfedresponse.htm#nrf.

[96] A summary of EPA responsibilities under various ESFs and HSPD-7 is provided in its factsheet "EPA's Support to the Water Sector Under the National Response Framework," available at http://www.epa.gov/safewater/watersecurity/pubs/fs_natl_response_frame work_nrf_fact_sheet.pdf.

[97] For information on the National Infrastructure Protection Plan and sector-specific agency roles, see the Department of Homeland Security's website, http://www.dhs.gov/national-infrastructure-protection-plan.

[98] 42 U.S.C. § 300i.

[99] For more information, see CRS Report RL34576, Managing Disaster Debris: Overview of Regulatory Requirements, Agency Roles, and Selected Challenges, by Linda Luther.

[100] Examples of EPA activities associated with post-disaster cleanup are available in CRS Report RL33115, Cleanup After Hurricane Katrina: Environmental Considerations, by Robert Esworthy et al..

[101] For more information, see CRS Report RL34576, Managing Disaster Debris: Overview of Regulatory Requirements, Agency Roles, and Selected Challenges, by Linda Luther.

[102] See EPA's website, "Fuels Waivers Response to 2005 Hurricanes," http://www.epa.gov /compliance/katrina/waiver/ index.html.

[103] This section is authored by Kelsi Bracmort, Specialist in Agricultural Conservation and Natural Resources Policy, Resources, Science, and Industry Division.

[104] The Department of the Interior (DOI) is the other federal entity, although DOI receives significantly less funding to manage wildfires.

[105] For more information on wildfire funding, see CRS Report RL33990, Federal Funding for Wildfire Control and Management, by Kelsi Bracmort.

[106] There are some instances where less aggressive suppression efforts are appropriate.

[107] NIFC is the national interagency support center for wildland firefighting.

[108] 42 U.S.C. §5187.

[109] For further analysis on SBA disaster loans see CRS Report R41309, The SBA Disaster Loan Program: Overview and Possible Issues for Congress, by Bruce R. Lindsay.

In: Federal Emergency Management
Editor: Lewis B. Adams

ISBN: 978-1-62618-217-2

Chapter 2

PRESIDENTIAL POLICY DIRECTIVE 8 AND THE NATIONAL PREPAREDNESS SYSTEM: BACKGROUND AND ISSUES FOR CONGRESS*

Jared T. Brown

SUMMARY

Presidential Policy Directive 8: National Preparedness (PPD-8) was signed and released by President Barack Obama on March 30, 2011. PPD-8 and its component policies intend to guide how the nation, from the federal level to private citizens, can "prevent, protect against, mitigate the effects of, respond to, and recover from those threats that pose the greatest risk to the security of the Nation." These threats include terrorist acts, natural disasters, and other man-made incidents. PPD-8 evolves from, and supersedes, Homeland Security Presidential Directive 8, which was released under President George W. Bush. PPD-8 is intended to meet many requirements of Subtitle C of the Post-Katrina Emergency Reform Act of 2006 (P.L. 109-295, 6 U.S.C. §741-764).

In addition to the main Directive, an Implementation Plan for PPD-8 and a National Preparedness Goal were finalized in 2011. Two National Planning Frameworks are also complete, but multiple component PPD-8

* This is an edited, reformatted and augmented version of the Congressional Research Service Publication, R42073, dated October 21, 2011.

policy documents are still being developed. Some elements of PPD-8 may not be finalized until September 2012 or later. However, PPD-8 has already affected national preparedness policy by expanding the scope of the end-state objective for preparedness, modifying the capabilities-based planning methodology, identifying a new set of national capabilities, and directing the creation of more National Planning Frameworks.

It is anticipated that the five National Planning Frameworks—one each for prevention, protection, mitigation, response, and recovery—will assign federal roles and responsibilities in each mission area. The National Planning Frameworks are also to guide how nonfederal resources are leveraged, including non-profit and private sectors' resources.

Congress may wish to oversee how the Administration creates and implements the many elements of PPD-8. This report discusses several potential issues and challenges that may arise in the development and implementation of each National Planning Framework. These issues and challenges include evaluating: how PPD-8 policies conform with statute; how federal roles and responsibilities have been assigned to implement and execute PPD-8 policies; how non-federal resources and stakeholders will be impacted by national preparedness guidance; and how the overall federal budget may be reprioritized by a new national preparedness goal. However, it may be difficult to ascertain the full impact of PPD-8 on national preparedness until its provisions are fully operationalized and tested during real world hazards.

OVERVIEW

The United States is threatened by a wide array of hazards, including catastrophic natural disasters, acts of terrorism, viral pandemics, and manmade disasters such as the Deep Horizon oil spill. The manner in which the nation prioritizes limited resources to prepare for disasters can significantly influence the ultimate cost to society, both in the number of human casualties and the scope of economic damage.[1]

The Post-Katrina Emergency Reform Act of 2006 (P.L. 109- 295, hereafter PKEMRA) mandates that the President develop a set of national policies to guide preparedness for these hazards, with the goal of reducing or preventing potentially devastating consequences.[2]

On March 30, 2011, President Barack Obama issued Presidential Policy Directive 8: National Preparedness (henceforth PPD-8) initiating the development of national preparedness policies that will fulfill many aspects of the mandate.[3]

Consistent with PKEMRA, the ultimate purpose of PPD-8 is to:

> strength[en] the security and resilience of the United States through systemic preparation for the threats that pose the greatest risk to the security of the Nation, including acts of terrorism, cyber attacks, pandemics, and catastrophic natural disasters.[4]

According to the Administration, this policy vision will be gradually implemented through a series of supporting documents to PPD-8, only a portion of which are currently finalized. Cumulatively, PPD-8 and its component policies are to guide how the nation "will prevent, protect against, mitigate the effects of, respond to, and recover from those threats that pose the greatest risk to the security of the Nation."[5] As required by PKEMRA, PPD-8 component policies include a National Preparedness Goal and a series of policies that are to collectively establish a National Preparedness System.

The practical impacts of PPD-8 and component policies are potentially far-reaching. PPD-8 policies could significantly influence the daily operations and priorities of homeland security officials, emergency managers, and first responders at the federal, state, and local government level. Further, PPD-8 policies may impact federal assistance and grant allocation decisions, guide the federal and nonfederal efforts to build homeland security and emergency management capabilities, and establish a national baseline for hazard preparedness. In future years, PPD-8 policies may also influence the Administration's budget requests and priorities.

Establishing national preparedness policy through a presidential directive has considerable precedent. PPD-8 rescinded the existing Homeland Security Presidential Directive 8: National Preparedness (henceforth HSPD-8),[6] which was released and signed by President George W. Bush on December 17, 2003. Prior to being rescinded, HSPD-8 and its supporting documents also instituted a strategic national preparedness policy that fulfilled many of the PKEMRA requirements, even before they were passed into law. Before HSPD-8, other administrations used directives to assign federal preparedness responsibilities, such as Executive Order 12656 issued under President Ronald Reagan.[7] PPD-8 policies are not considered a repudiation of HSPD-8 policies; rather they are viewed as a policy evolution from the chain of national preparedness policies that has preceded it. PPD-8 policy evolves from HSPD-8 by:

- Expanding the scope of the end-state objective in the National Preparedness Goal to include mitigation as a key mission area for

national preparedness, in addition to prevention, protection, response, and recovery;
- Modifying the capabilities-based planning approach through a reduced focus on national planning scenarios;
- Identifying a different set of national capabilities needed to achieve national preparedness in each mission area; and
- Directing the creation of National Planning Frameworks for all identified mission areas of national preparedness.

This report explains the importance of this evolution, and provides a summary of the implementation status of the directed activities of PPD-8. Further, the report discusses the possible content of the future National Planning Frameworks being produced by the Executive Branch pursuant to PPD-8 requirements. Finally, the report provides a summary of some of the issues Congress may wish to consider when it oversees the development and implementation of national preparedness policy. These include evaluating: how PPD-8 policies conform with PKEMRA statute; how federal roles and responsibilities have been assigned to implement and execute PPD-8 policies; how non-federal resources and stakeholders will be impacted by national preparedness guidance; and how the overall federal budget may be reprioritized by a new national preparedness goal.

BACKGROUND ON KEY CONCEPTS

National Versus Federal Preparedness Policy

The Obama Administration describes a "Presidential Policy Directive" as a mechanism "for communicating presidential decisions about national security policies of the United States."[8] The Department of Justice's Office of Legal Counsel has opined that there is no substantive legal difference between a presidential directive, such as PPD-8, and an executive order.[9] Though the legal authority of Executive Orders is vaguely defined, they have been used by all Presidents for a variety of purposes, including to delegate authorities granted to the President by Congress, to achieve policy goals, set uniform standards for managing the executive branch, or outline a policy view intended to influence the behavior of private citizens.[10.] In this instance, the authority of the presidential directive is used to order elements of the federal government to develop a set of national preparedness policies. Further, PPD-8 component policies will assign responsibilities to federal agencies to execute the necessary national preparedness capabilities (namely, in the National Planning Frameworks discussed later). However, the President does not have the authority to direct the resources and authorities of state and local governments, the private sector, the non-profit sector, and normal citizens through PPD8 and subsequent documents.[11] Traditionally, however, stakeholders in emergency management and national security have generally adhered to the national security guidance provided in presidential directives.[12] Therefore, PPD-8 policy is intended not just to establish a federal set of policies and plans on national preparedness, it is also aimed at "establishing common intent and fostering robust partnerships across all communities and levels of government; building the capacity of partners across jurisdictional boundaries; and encouraging dynamic coordination and cooperation" in national preparedness activities.[13] With these dual objectives, PPD-8 policies may establish *national* preparedness policy and plans, not just federal policy and plans. As a national policy, PPD-8 and its components are core elements of the Administration's overarching National Security Strategy.[14]

Capability-Based Planning for All Hazard Preparedness

A capability is defined in PKEMRA as "the ability to provide the means to accomplish one or more tasks under specific conditions and to specific performance standards. A capability may be achieved with any combination of properly planned, organized, equipped, trained, and exercised personnel that achieves the intended outcome."[15] This definition is echoed in the Implementation Plan for PPD-8.[16] Both PPD-8 and HSPD-8 rely on what is generally referenced to as a "capability-based planning" approach to develop national preparedness policy. The capability- based planning concept in homeland security evolved from traditional national defense planning.[17] No universally accepted definition exists for capability-based planning, but Paul K. Davis of the RAND National Defense Research Institute has defined it in the context of Department of Defense planning. Davis writes:

> Capabilities-based planning is planning, under uncertainty, to provide capabilities suitable for a wide range of modern-day challenges and circumstances, while working within an economic framework. This seemingly innocuous definition has three important features. First, the

> notion of planning under uncertainty appears in the very first clause: uncertainty is fundamental, not a mere annoyance to be swept under the rug. Second, the idea is to develop capabilities—i.e., the general potential or wherewithal—to deal effectively not just with a well defined single problem, but with a host of potential challenges and circumstances... Third, this is to be done not with the largesse of a blank-check policy (preparing for anything that might conceivably arise), but rather while working within an economic framework.[18]

FEMA has created a similar definition in the context of emergency management. FEMA writes:

> Planning, under uncertainty, to provide capabilities suitable for a wide range of threats and hazards while working within an economic framework that necessitates prioritization and choice. Capabilities-based planning addresses uncertainty by analyzing a wide range of scenarios to identify required capabilities.[19]

In theory, an advantage of capability-based planning is that it is scalable and flexible enough to meet the unique and unanticipated challenges of an unknown set of threats and disasters. The use of capability-based planning, however, does not preclude the use of specific scenario plans. For example, the HSPD-8 planning process included scenario planning into the approach and released several specific disaster scenario plans, but this was mainly done as a means of identifying the range of possible capabilities required for the nation to be safe. However, if the planning becomes too focused on specific scenarios, Davis notes that the plans often become "characterized by a fixation on particular enemies, particular wars, and particular assumptions about those wars—a fixation that comes at the expense of more flexible and adaptive planning."[20]

Given the theoretical flexibility of capability-based planning, it is an apt planning method for developing "all hazard" preparedness policy. The Robert T. Stafford Disaster Relief and Emergency Assistance Act (P.L. 93-28 8, hereafter the Stafford Act) defines "hazard" as any emergency or disaster resulting from a natural disaster or an accidental or man-caused event.[21] PKEMRA adopts the same definition, and stipulates further that the national preparedness system should be established "in order to prepare the Nation for all hazards, including natural disasters, acts of terrorism, and other man-made disasters...."[22] Despite the broad inclusiveness of "all hazard" preparedness policy, many types of hazards may not be fully accounted for under this policy. For instance, there are "slow-onset" natural hazards[23] that will likely go unaddressed by the policies of PPD-8, as their time horizon is better handled through traditional land-use, urban development, environmental, or economic policy.[24] Further, though included under the hazard definition, "accidental" or technological disasters may be beyond the scope of some mission areas of national preparedness policy, especially in preventing and protecting against those hazards. Disasters like oil spills, electrical blackouts, train derailments, and dam collapses that are caused by human error or technological failures are often prevented and protected against through government and private-sector regulatory and safety procedures. Therefore, it is likely that PPD-8 policies will guide how the nation responds to and recovers from technological hazards, but it is unlikely that PPD-8 policies would encompass or replace the established procedures for preventing and protecting against these types of hazards.

Current Status of PPD-8 Implementation

In addition to the main Directive, an Implementation Plan for PPD-8 and the National Preparedness Goal were completed in 2011. Two of the National Planning Frameworks are also complete.[25] Multiple component PPD-8 policy documents are still being developed. Table 1 summarizes the purpose and current status of the PPD-8 supporting documents.

National Preparedness Goal

During the development of the National Preparedness Goal, FEMA solicited public comments on a draft version in late August, early September of 2011.[26] The draft was then revised, and the final version of the Goal was released in late September. The end-state objective for national preparedness, as identified in the finalized National Preparedness Goal, is:

> A secure and resilient Nation with the capabilities required across the whole community to prevent, protect against, mitigate, respond to, and recover from the threats and hazards that pose the greatest risk.[27]

The end state objective in PPD-8 does not dramatically differ from that of HSPD-8.[28] However, PPD-8 notably differs from HSPD-8 in that it addresses mitigating threats and hazards.[29] This places added emphasis on reducing the consequence of potential disasters, both before and after an incident, by increasing the nation's overall resilience to adversity. As defined in the Directive, mitigation includes capabilities both before and after a disaster, that transcend the preparedness spectrum.[30] On one hand, some might argue that having a separate mission area for mitigation will provide heightened awareness of its importance in each of the other mission areas, leading to a nation that is efficiently mitigating all hazards. Conversely, one could argue that isolating mitigation capabilities in a separate mission may lead those capabilities to being ignored or overlooked by the stakeholders of the other mission areas, even though they are capabilities that should be employed in all phases of the preparedness cycle.

To understand the conceptual importance of this shift, it may be useful to think of the process of building a house. The objective statement of the National Preparedness Goal represents the architect's drawing of the house. At this stage, the drawing is scant in detail, but it illustrates the architect's general vision for how the house will look when completed. With the inclusion of mitigation as an element of the national preparedness goal, PPD-8 has added something akin to a new floor in the architectural drawing. It is hard to assess the value of adding this floor to the design at this point in time, but it is worth noting as a distinct evolution from HSPD-8.

In addition to defining the end-state objective, the National Preparedness Goal also identifies the set of core capabilities needed to reach the end-state objective in each of the mission areas of prevention, protection, mitigation, response, and recovery. The Goal also identifies preliminary target levels for these capabilities. Through a capability-based planning process, 32 different core capabilities were identified across the five mission areas of the National Preparedness Goal. Three core capabilities were identified as being required in all mission areas.

The core capabilities are listed in table 2 of the National Preparedness Goal (also provided in Table 2 of this report). By comparison, the HSPD-8 Target Capability List had five common capabilities and 29 capabilities across four mission areas. The National Preparedness Goal suggests that FEMA will provide a "cross-walk" of the two lists on its website, but one has yet to be provided.[31] One example of a PPD-8 core capability is "Access Control and Identity Verification" in the protection mission area. In order to protect the nation against all hazards, the nation must be able to "Apply a broad range of physical, technological, and cyber measures to control admittance to critical locations and systems, limiting access to authorized individuals to carry out legitimate activities."[32]

Table 1. Status of PPD-8 Component Policy Documents

Title	Purpose	Status	Related PKEMRA Requirement
Implementation Plan to PPD-8	Guides the execution of PPD-8 by expanding upon the purpose of the required directed actions. Assigns responsibilities in the Executive Branch to develop the elements of the Directive. Sets deadlines for steps in the implementation process and the issuance of supporting policy documents to PPD-8.	Completed, May 2011	N/A
National Preparedness Goal	Provides the end-state objective for national preparedness. Identifies the required capabilities to achieve the end-state goal.	Completed, September 2011	6 U.S.C. §743
National Preparedness System Description	Will identify and describe all elements of the national preparedness system that are needed to execute the required capabilities.	Due November 24, 2011	6 U.S.C. §744
National Preparedness Report	Provides an annual report on the progress of the nation in achieving the National Preparedness Goal. The Report will be provided the President and appropriate committees of Congress. Acts as a "status report" on where the Executive Branch is in developing the needed capabilities in fulfillment of PKEMRA reporting requirements.	Due March 30, 2012	6 U.S.C. §752
National Planning Frameworks	Assigns key roles and responsibilities to Executive Branch agencies to provide the necessary capabilities for each mission area. Develops plans for leveraging the resources and authorities of states, local governments, the private and non-profit sectors, and the public in each mission area.	Due June 30, 2012	
• National Prevention Framework	Identifies capabilities necessary to avoid, prevent, or stop a threatened or actual act of terrorism. These capabilities include, but not limited to, information sharing and warning; domestic counterterrorism activities; and preventing the acquisition or use of weapons of mass destruction (WMD).	Due June 30, 2012	N/A

Table 1. (Continued)

Title	Purpose	Status	Related PKEMRA Requirement
• National Protection Framework	Identifies capabilities necessary to secure the nation against acts of terrorism, and manmade or natural disasters. These capabilities include, but not limited to, defense against WMD threats; defense of agriculture and food threats; critical infrastructure protection; key leadership and events protection; border security; maritime security; transportation security; immigration security; and cybersecurity.	Due June 30, 2012	N/A
• National Mitigation Framework	Identifies capabilities necessary to reduce loss of life and property by lessening the impact of disasters. These capabilities include, but not limited to, community-wide risk reduction projects; improving resilience of critical infrastructure and key resources; risk reduction for specific vulnerabilities from natural disasters and acts of terrorism; and initiatives to reduce future risks after a disaster.	Due June 30, 2012	N/A
• National Response Framework	Identifies capabilities necessary to save lives, protect property and the environment, and meet basic human needs after a disaster.	Existing edition completed January 2008	6 U.S.C. §319
• National Disaster Recovery Framework	Identifies capabilities necessary to assist communities affected by a disaster to effectively recover, including, but not limited to, rebuilding infrastructure; providing adequate interim and long-term housing; restoring health, social, and community services; promoting economic development; and restoring natural and cultural resources.	Completed, September 2011	6 U.S.C. §771
Interagency Operational Plans	Accompany each National Planning Framework. Include a detailed concept of operations (CONOPs), detailed task lists, and equipment/resource requirements.	September 25, 2012	6 U.S.C. §753(b)

Source: P.L. 109-295, PPD-8, the Implementation Plan for PPD-8, and the National Preparedness Goal.

PPD-8 also evolved from HSPD-8 by altering the process used to identify the set of national capabilities. Both HSPD-8 and PPD-8 used a capability-based planning approach, but HSPD-8 relied more heavily on a set of national scenarios to derive the capabilities necessary for national preparedness during the planning process. PKEMRA states that the President "may" include National Planning Scenarios in development of preparedness policy, but does not explicitly require their usage.[33] Under HSPD-8, 15 different National Planning Scenarios were developed that helped identify the capabilities needed for national preparedness.[34] In contrast, PPD-8 policy does not appear to use Planning Scenarios extensively to help identify capabilities for the Goal, nor is there mention that scenarios will be used in the future. In the draft of the PPD-8 National Preparedness Goal, there was mention of a "meta-scenario" that was used to define the capabilities needed for response and recovery.[35] This meta-scenario was not mentioned specifically in the final version of the Goal, and appears to have been simplified in text to a "no- notice, cascading incident".[36] However, the meta-scenario may still have been used as a template for Administration officials to develop the capability list.

In addition to the possible use of the meta-scenario, a "Strategic National Risk Assessment" was completed to help identify capabilities. In PPD-8, the President mandated that

> The Secretary of Homeland Security will conduct a strategic, national-level risk assessment to identify the relevant risk factors that guide where core capabilities are needed and develop a list of the capabilities and associated performance objectives for all hazards that will measure progress toward their achievement. The national risk assessment should build on and integrate current models and best practices to enable the national assessment to be applied regionally and on a local level, as appropriate and practicable. The Director of National Intelligence and Attorney General will provide relevant and appropriate terrorism- related intelligence information to the Secretary of Homeland Security for the development of the risk assessment.

An extensive description of the risk methodology used in this assessment is not publically available.[37] Also, the complete results of the Assessment are classified.[38] However, the Goal mentions several broad key findings, such as that "Cyber attacks can have their own catastrophic consequences and can also initiate other hazards, such as power grid failures or financial system failures, which magnify the potential impact of cyber incidents."[39] It is difficult to assess the value of these findings in identifying necessary capabilities without access to the classified version of the Assessment. However, on the surface, these findings appear to be too broad for them to be substantially helpful in identifying capabilities needed in any of the mission areas.

Returning to the metaphor of the house, the process used by National Preparedness Goal to identify capabilities is like the development of a engineering blue print from the architect's drawing of that house. The engineering blue print adds considerable detail to the plan by identifying the core features of the house, such as the plumbing, the electrical wiring, and the masonry walls. Therefore, in assessing the blueprint, the construction foreman knows that in order to build the house, he will need the capability to install plumbing, construct masonry structures, or safely install electrical wiring. The provided example capability, "Access Control and Identity Verification," is like the capability to install plumbing—it is a necessary set of activities that the nation needs to be able to perform successfully in order to reach the end-state objective of a secure and resilient nation. At this point in the process, the Goal has identified what type of house the administration would like to build, its key features, and the needed capabilities to construct the house. It has yet to identify who, how, and with what tools and resources, these capabilities will be accomplished. That part of the policy is developed in the National Preparedness System, namely through the National Planning Frameworks and companion interagency operational plans.

National Planning Frameworks

The National Preparedness System has numerous elements, but a series of five National Planning Frameworks will arguably be most significant for overall preparedness policy. The Frameworks, one for each mission area of national preparedness, are to assign key roles and responsibilities to federal agencies and help determine how limited resources and authorities are applied to develop national preparedness. HSPD-8 only specifically called for one document, the National Response Plan, which is the predecessor to the National Response Framework. PPD-8 states:

> "The frameworks shall be built upon scalable, flexible, and adaptable coordinating structures to align key roles and responsibilities to deliver the necessary capabilities. The frameworks shall be coordinated under a unified system with a common terminology and approach, built around basic plans that support the all-hazards approach to preparedness and

> functional or incident annexes to describe any unique requirements for particular threats or scenarios, as needed. Each framework shall describe how actions taken in the framework are coordinated with relevant actions described in the other frameworks across the preparedness spectrum."[40]

Unlike HSPD-8, the key terms prevention, protection, mitigation, response, and recovery have been defined in PPD-8. In the Implementation Plan, each Framework has also been assigned to an Interagency Policy Committee (IPC) of the National Security Council (NSC).[41]

The assigned IPC is to oversee the policy coordination of the Frameworks, and provide a forum for resolving policy disputes that may arise in their development. The lead federal government officials have also been designated for each Framework in the Implementation Plan. The National Preparedness Goal lists the capabilities of each Framework mission area, and elaborates further on the definitions of the mission area.[42] This information is summarized in Table 2 of the report.

Though PPD-8 describes the Frameworks as being part of a cycle of preparedness, the capabilities covered under each Framework are not necessarily aligned in a chronological order. Prevention capabilities do not always precede protection capabilities, protection does not always precede response, and so forth.

For example, the National Prevention Framework is to cover capabilities that may be necessary both before and after the traditional response phase to a disaster.[43] In particular, the National Mitigation Framework encompasses capabilities that are necessary in all four other focus areas, as it deals with the long-term activities to reduce hazard risk and vulnerability by increasing the nation's resilience. Further, three core capabilities are required under all of the Frameworks.[44]

In order to achieve the core capabilities, the Frameworks are not intended to rely solely on the appropriated resources of the federal government, or the authority from any one statute.

As discussed earlier, PPD-8 has the authority to direct federal authorities and resources in the provision of the capabilities, but can only provide guidance to the rest of the nation. Through this national guidance, PPD-8 aims to leverage national resources and authorities from state and local governments, private and nonprofit sectors, and the public for the delivery of these core capabilities. At a federal level, leveraged authorities and resources may come from a variety of legislative sources.[45]

Table 2. Summary of the National Planning Frameworks

Title	Interagency Policy Committee [a]	Federal Leadership[a]	Core Capabilities [b]	Definition of Mission Area [c]
National Prevention Framework	Counterterrorism Security Group	Attorney General, Secretary of Homeland Security, and Director of National Intelligence	Forensics & Attribution; Intelligence & Information Sharing; Interdiction & Disruption; Search & Detection.	The term "prevention" refers to those capabilities necessary to avoid, prevent, or stop a threatened or actual act of terrorism. Prevention capabilities include, but are not limited to, information sharing and warning; domestic counterterrorism; and preventing the acquisition or use of weapons of mass destruction (WMD). For purposes of the prevention Framework called for in this directive, the term "prevention" refers to preventing imminent threats.[d]
National Protection Framework	Transborder Security Group	Secretary of Homeland Security[e]	Access Control and Identity Verification; Cyber-security; Intelligence & Information Sharing; Interdiction & Disruption; Physical Protective Measures; Risk Management for Protection Programs and Activities; Screening, Search, and Detection; Supply Chain Integrity and Security.	The term "protection" refers to those capabilities necessary to secure the homeland against acts of terrorism and manmade or natural disasters. Protection capabilities include, but are not limited to, defense against WMD threats; defense of agriculture and food; critical infrastructure protection; protection of key leadership and events; border security; maritime security; transportation security; immigration security; and cybersecurity.

Title	Interagency Policy Committee [a]	Federal Leadership[a]	Core Capabilities [b]	Definition of Mission Area [c]
National Mitigation Framework	Domestic Resilience Group	Secretary of Homeland Security	Community Resilience; Long-Term Vulnerability Reduction; Risk and Disaster Resilience Assessment; Threats and Hazard Identification.	The term "mitigation" refers to those capabilities necessary to reduce loss of life and property by lessening the impact of disasters. Mitigation capabilities include, but are not limited to, community- wide risk reduction projects; efforts to improve the resilience of critical infrastructure and key resource life lines; risk reduction for specific vulnerabilities from natural hazards or acts of terrorism; and initiatives to reduce future risks after a disaster has occurred.
National Response Framework	Domestic Resilience Group	Secretary of Homeland Security	Critical Transportation; Environmental Response/ Health and Safety; Fatality Management Services; Infrastructure Systems; Mass Care Services; Mass Search and Rescue Operations; On-Scene Security and Protection; Operational Communication; Public and Private Services and Resources; Situational Assessment.	The term "response" refers to those capabilities necessary to save lives, protect property and the environment, and meet basic human needs after an incident has occurred.[d]
National Disaster	Domestic Resilience Group	Secretary of Homeland Security,	Economic Recovery; Health and Social Services; Housing;	The term "recovery" refers to those capabilities necessary to assist communities affected by an incident to

Table 2. (Continued)

Title	Interagency Policy Committee [a]	Federal Leadership[a]	Core Capabilities [b]	Definition of Mission Area [c]
Recovery Framework		Secretary of Housing and Urban Development	Infrastructure Systems; Natural and Cultural Resources Recovery	recover effectively, including, but not limited to, rebuilding infrastructure systems; providing adequate interim and long-term housing for survivors; restoring health, social, and community services; promoting economic development; and restoring natural and cultural resources.

Sources: White House, *Presidential Policy Directive 8: National Preparedness*, Washington, DC, March 30, 2011, http://www.dhs.gov/about/laws White House, *Implementation Plan for Presidential Policy Directive 8: National Preparedness*, Washington, DC, May 2011; Federal Emergency Management Agency, *National Preparedness Goal*, National Review Draft, Washington, DC, August 2011, http://www.fema.gov/pdf/prepared/npg.pdf.

a. This information is found on pp. 8-10 of the Implementation Plan.

b. This information is found on p. 2 of the National Preparedness Goal. Planning, Public Information and Warning, and Operational Coordination are capabilities needed in all mission areas.

c. These definitions are found on p. 6 of the PPD-8. Additional explanation of the mission areas definitions are provided on p. 4-18 of the National Preparedness Goal.

d. Page 9 of the *Implementation Plan* draws a further distinction between the terms response and prevention. It states: " 'Response' will not include capabilities related to law enforcement investigative and intelligence response to a threatened or actual terrorism event, which are covered under Prevention and in the National Prevention Framework. The prevention and response Frameworks will include language describing how investigation and intelligence actions taken during a threatened or actual terrorism event align with response actions taken under the National Response Framework."

e. Except defense activities, which are the responsibility of the Secretary of Defense. This will likely involve the protection of the defense industrial base.

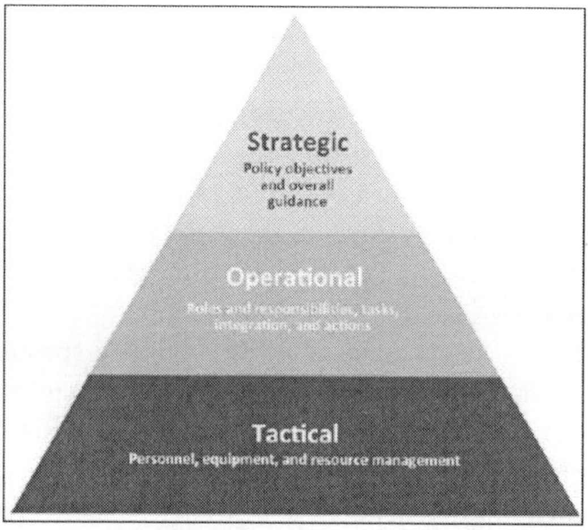

Source: Federal Emergency Management Agency, Comprehensive Preparedness Guide 101: Developing and Maintaining Emergency Operations Plans Version 2.0, Washington, DC, November 2010, pp. 1-5, http://www.fema.gov/pdf/about/divisions/npd/CPG_101_V2.pdf.

Figure 1. Relationship between Strategic, Operational, and Tactical Planning.

Ultimately, these Frameworks may help operationalize the Directive's strategic guidance for the majority of stakeholders in federal, state, and local emergency management and homeland security agencies. Each Framework is to be accompanied by an interagency operational plan that includes "a more detailed concept of operations; description of critical tasks and responsibilities; detailed resource, personnel, and sourcing requirements; and specific provisions for the rapid integration of resources and personnel."[46] Though the Frameworks themselves serve as a guide for the whole nation, it is anticipated that the interagency operational plan will only be prescriptive to the federal government.

Returning again to the homebuilding metaphor, one can think of the national authorities and resources leveraged by the Frameworks as the very large set of specialized construction equipment and tools that are required to build a house. Some of the tools are fundamental to the construction project, such as a hammer, just like many authorities, such as the Stafford Act, will be fundamental to many of the Frameworks. The interagency operational plans are analogous to basic instructions on how to use the equipment for the execution of certain capabilities in the building of the house. Another way of conceptualizing the hierarchy of national preparedness policies is through a commonly used pyramid structure.

Figure 1 displays the difference between levels of preparedness planning. For national preparedness policy, the strategic level of planning is provided by the main PPD-8 document, the Implementation Plan, and the National Preparedness Goal. The National Planning Frameworks and the other elements of the National Preparedness System are to establish the operational level of the planning pyramid, and they begin to shape the tactical level of planning for the federal government in the accompanying interagency operational plans. However, the majority of tactical level planning would be beyond the scope of any PPD-8 policy, as it necessitates level of detail and specificity that cannot be generalized nationally. In a sense, the National Planning Frameworks are to translate the strategic vision of the National Preparedness Goal into a more accessible, though detailed, set of directions for stakeholders on how the nation should prepare for disasters.

Content of the Frameworks[47]

Table 2 summarizes the key information available on each Framework called for under PPD-8, its Implementation Plan, and the National Preparedness Goal. Using this information as a baseline, this section of the report discusses the content of the Frameworks, and identifies potential policy issues and challenges associated with their development and content. The Implementation Plan stipulates that the Frameworks should "build on and leverage established relevant planning documents wherever possible."[48] The National Disaster Recovery Framework (NDRF) has already been released under PPD-8, and its content and structure may help predict the remaining Frameworks.[49] Two other Frameworks have existing policy documents that can serve as precedent. The National Response Framework (NRF) will probably remain as is or be updated from its current version,[50] and the National Infrastructure Protection Plan (NIPP) will likely serve as a baseline document for the creation of the National Protection Framework.[51] The remaining documents, the National Prevention Framework and the National Mitigation Framework, do not have equally significant precedent, and therefore their content is more difficult to predict.

Coordinating Structure and the Support Function Model

PPD-8 states that the Frameworks "shall be built upon scalable, flexible, and adaptable coordinating structures to align key roles and responsibilities to deliver the necessary capabilities."[52] Extrapolating from existing precedent, there is an existing coordinating structure that could be adopted in all of the Frameworks—the Support Function model used by the NRF and NDRF. There are other alternatives, such as the Sector Specific Agency coordinating model of the NIPP,[53] but the Support Function model is most applicable to the capability-based planning approach of the Directive. No matter what coordinating structure is used to align responsibilities, it may be applied consistently throughout all the Frameworks, in adherence to the policy guidance that all of the Frameworks should be "coordinated under a unified system with a common terminology and approach."[54]

Therefore, it is anticipated that the NRF and NDRF support function structure will be adopted in the remaining Frameworks. Support functions are essentially teams of federal agencies working together to accomplish a set of core capabilities in a particular mission area. Returning to the house metaphor, one can think of a support function as the team of plumbers who are assigned to install the plumbing for the house. For example, in the NRF, Emergency Support Function (ESF) #9 is "Search and Rescue" and it unifies federal agencies with the appropriate resources and authorities to conduct search and rescue operations following a hurricane, earthquake, or other disaster. Similarly, the Recovery Support Function (RSF) "Infrastructure Systems" of the NDRF will create a team of federal agencies with appropriate resources and authorities in order to provide the core capability of "Infrastructure Systems" in the recovery mission.

Support functions typically have one coordinating agency (often imprecisely referenced to as the "lead" agency), several primary agencies, and many more support agencies; each with a decreasing amount of responsibility and relevant authorities, resources, and expertise relating to the core capabilities of the support function.[55] If the model of the NDRF is followed in the remaining Frameworks, there may be a support function paired to each capability. However, this is not the model adopted under the current version of the NRF, which was released prior to PPD-8.

National Prevention Framework

Unlike the other Frameworks that focus on all potential sources of major hazards, the National Prevention Framework is to focus only on preventing an imminent terrorist threat in the homeland. This Framework may discuss how the nation will "connect the dots" of terrorism intelligence, and how actionable intelligence would be shared with stakeholders in order to prevent or disrupt these imminent threats. It may also cover how the nation would detect and interdict "sleeper cells" or "lone wolf" terrorists before they initiate a terrorist attack. It is likely that the Framework will leverage the current policies of the information sharing environment, and existing prevention resources such as the National Counterterrorism Center and associated fusion centers.[56] In the event of a foiled or successful terrorist attack, the Framework could also address the capabilities needed to investigate and find the persons responsible for the attack. Because of its restriction to imminent threats, it is unlikely to deal with longer-term issues such as countering violent extremism.[57]

Potential Issues and Challenges

- Counterterrorism efforts require close coordination between domestic and international security activities. What impact will the Framework have on international efforts to prevent terrorism by the State Department, Department of Defense, and national intelligence agencies? How will the Framework be coordinated with these existing counterterrorism and intelligence efforts?
- There are a multitude of federal agencies, especially within the Departments of Homeland Security and Justice, with intelligence and law enforcement authorities and resources.[58] It may be particularly challenging to coordinate their available resources and assign responsibilities in this Framework. For instance, how will the roles and responsibilities of the Director of National Intelligence or the Director of the Central Intelligence Agency be delineated from the Attorney General and Director of the Federal Bureau of Investigation, or from the DHS Secretary and Undersecretary of Intelligence and Analysis?

National Protection Framework

The National Protection Framework is to encompass a wide range of homeland security topics, including critical infrastructure and key resource security and protection; border and immigration security; supply chain security; cyber security; the detection and interdiction of weapons of mass destruction; and continuity of operations/government.

Arguably, it is the Framework with the broadest portfolio of responsibilities, though this is not reflected in a simple numeric count of assigned core capabilities. The National Infrastructure Protection Plan (NIPP) may serve as a baseline for the infrastructure security and protection issues, but numerous other policy documents and initiatives will probably be leveraged in the development of the Framework.[59]

Potential Issues and Challenges

- Can all the potential topics of the Framework, and their associated resources and authorities, be reasonably addressed in a single scalable, flexible coordinating structure? Will relatively "small" topics like continuity of operations/government be isolated or overlooked given the immense scope of the Framework?
- Can the focus of the Framework be appropriately shared across the spectrum of natural and human-made hazards? Will traditional protection measures against natural hazards, such as reinforced bridges for earthquakes or levee systems for floods, be neglected if the focus of the Framework shifts to protecting the border, cyber security, and transportation security? Or vice versa?
- Will the Framework create risk-based priorities for federal protection resources, and if so, will these priorities be reflected in preparedness grant programs targeted at prevention and in other elements of the National Preparedness System?

National Mitigation Framework

The content of the National Mitigation Framework may be the most difficult to predict. As defined in PPD-8, mitigation includes capabilities both before and after a disaster.[60] The National Preparedness Goal further asserts that "Spanning across community planning, critical infrastructure, public health, and future land use, mitigation requires an understanding of the threats and hazards that, in turn, feed into the assessment of risk and disaster resilience in the community."[61] FEMA has been the lead agency working in the traditional field of hazard mitigation.[62] However, the policies advocated in PPD-8 and Goal appear to widen the scope of traditional hazard mitigation to include land use, environmental, infrastructure, and health policy. This expansion could lead to the extensive involvement of non-traditional partners, authorities, and resources that are typically associated with "normal" rural and urban development.

Potential Issues and Challenges

- A majority of states already have hazard mitigation plans, as a result of FEMA grants and financial incentives.[63] How will the Framework incorporate the consistent elements of these mitigation plans? Will the creation of the Framework impact the design of the state plans, or necessitate their revision? If so, how?

- In floodplain management, the nation attempts to mitigate flood damage through the construction of levees to protect areas within certain zones of risk, namely areas with 1%-annual-chance-of-flooding.[64] What levels of risk will the National Mitigation Framework use to dictate mitigation measures for other threats and hazards? Will the 1%-annual-chance standard be applied to mitigating terrorist events, or earthquakes, or pandemics as well?

National Response Framework

The National Response Framework (NRF) was finalized in 2008.[65] The NRF is the latest iteration of an evolutionary chain of policy documents that guided how the nation responded in the immediate aftermath of an incident.[66] The Administration may determine that the current NRF is sufficient to fulfill the mission detailed under PPD-8, but there are two major reasons why it could be revised instead. First, it may be necessary to develop protocols on how response operations are transitioned to recovery operations of the NDRF, as well as how the NRF response capabilities relate with the other Frameworks.[67] Second, the draft version of National Preparedness Goal also discussed a "meta-scenario" that was potentially used to drive the development of response and recovery capabilities.[68] The extreme catastrophic nature of the meta-scenario may necessitate revision of the NRF simply because the event demands the nation to achieve higher standards of response capabilities.

This higher standard may led to a realignment of ESF roles and responsibilities, a further expansion of the existing ESF structure, or a more formal incorporation of the concepts in the NRF 's Catastrophic Incident Annex[69] in a newly revised NRF.

Potential Issues and Challenges

- The NRF, and its ESF structure, has been adopted by significant majority of states, and many local governments, to guide their own disaster response efforts. If the NRF and its ESF structure are revised, how will the changes to the Framework be communicated to stakeholders, and once communicated, will the changes be adopted? If they aren't adopted, how would the potential ensuing confusion or conflict between old and new NRF policies be overcome?

- There have been recent concerns over the cost of federal disaster assistance and the increasing number of disaster declarations through the Stafford Act. If revised, does the NRF increase or decrease federal government's responsibilities to provide disaster response assistance, both technically and financially?
- Response, recovery, and mitigation capabilities often overlap following a disaster. Who is responsible for coordinating national efforts across the Frameworks?

National Disaster Recovery Framework

The NDRF is first national Framework released under PPD-8.[70] The development of the NDRF began well before the release of Directive in March of 2011. The concept of a National Strategy for Disaster Recovery was first mandated by Section 682 of PKEMRA.[71]

In September of 2009, the President requested that the Secretaries of DHS and Housing and Urban Development (HUD) co-chair a "White House Long-Term Disaster Recovery Working Group" to develop this national policy on disaster recovery.[72]

In addition to the NDRF, the Working Group was charged with completing a "Report to the President" that would identify opportunities for improving the delivery of future recovery assistance.[73] The accompanying "Report to the President" has yet to be released.

The NDRF uses the Support Function coordinating structure. Instead of the fifteen ESFs of the NRF, the NDRF establishes six Recovery Support Functions (RSFs).[74]

Reflecting the Federal Coordinating Officer (FCO) position of the NRF, the NDRF also establishes a Federal Disaster Recovery Coordinator (FDRC) position to lead Federal interagency efforts.[75] Notably, FEMA is the coordinator for only one of the six RSFs.

Potential Issues and Challenges

- Does the NDRF or Report to the President meet all requirements of PKEMRA, while also fulfilling the core capabilities of the Framework as outlined in the National Preparedness Goal? The Working Group, through FEMA, released a draft version of the NDRF in the Federal Register (FR) on February 10, 2010.[76] This draft was broadly criticized in a letter supplied for public comment by the Senate

Committee on Homeland Security and Government Affairs of the 111th Congress.[77] The final version of the NDRF does not resolve many of the issues suggested in the letter. The final NDRF fails to:

- Meet the PKEMRA requirement to outline the most efficient and cost- effective federal programs that will meet recovery needs;[78]
- Describe in detail any funding issues in the federal programs that may be offered as assistance;[79] or
- Promote the Voluntary Private Sector Preparedness Accreditation and Certification Program (PS-Prep) program, as advocated in the Senate letter.[80]

- One of the major concerns raised during the development of the NDRF was that present authorities and resources may be too limited to enable a resilient disaster recovery process, especially after a catastrophic scenario. Are the federal government's current authorities and resources sufficient to meet the challenge of recovering from the draft National Preparedness Goal's "meta-scenario"? Is new legislation required to further enhance the capability of the federal government?

ADDITIONAL COMPONENTS OF THE NATIONAL PREPAREDNESS SYSTEM

There are other components of the National Preparedness System that are mentioned in the strategic guidance of PPD-8, the Implementation Plan, and the National Preparedness Goal.

Currently, there is little information available on the additional elements, preventing detailed analysis. These elements will likely be described further in a forthcoming description of the National Preparedness System, due on November 24, 2011.[81] Table 3 summarizes what is known about these elements and how they are possibly related PKEMRA requirements.

Table 3. Additional Elements of the National Preparedness System

Title	Purpose	Possible Related PKEMRA Requirement(s)	Notes
Campaign to Build and Sustain Preparedness	"... will provide an integrating structure for new and existing community-based, nonprofit, and private sector preparedness programs, research and development activities, and preparedness assistance" [a]	6 U.S.C. §753	The Campaign may include public outreach, federal assistance to develop national preparedness (such as grant programs), and a research and development campaign. P. 6 of the Implementation Plan explains these elements further.
National Training and Education System	"... to provide a structure for the development of the core capabilities identified to meet the Goal. Training will be targeted to the specific skill sets required to meet each capability's performance objectives, organized around learning objectives that identify the tasks that need to be performed."[b]	6 U.S.C. §747(b)	This may incorporate FEMA's existing training and education programs, primarily through the Emergency Management Institute[c].
National Exercise Program	"Provides a Framework for prioritizing and coordinating federal, regional and state exercise activities, without replacing any individual department or agency exercises. The NEP enables federal, state and local departments and agencies to align their exercise programs." [d]	6 U.S.C. §748	A National Exercise Program is already established under HSPD-8 guidance. It is unclear how PPD-8 might revise the existing program. [e]
Remedial Action Management Program	"... to collect, share, and track progress on lessons learned and corrective actions undertaken to build and sustain preparedness"[f]	6 U.S.C. §749 and 750	

Source: : P.L. 109-295, PPD-8, the Implementation Plan for PPD-8, and the National Preparedness Goal.

a. National Preparedness Goal, p. 19.

b. Implementation Plan, p. 6.

c. See http://training

d. See the existing National Exercise Program's website at http://www.dhs.gov/files/ training

e. For more, see CRS Report RL34737, *Homeland Emergency Preparedness and the National Exercise Program: Background, Policy Implications, and Issues for Congress*, coordinated by R. Eric Petersen.

f. Implementation Plan, p. 3.

Issues for Congress

Conformity with PKEMRA

One of the major issues Congress may wish to consider is the conformity of PPD-8 and component policies with the full intention of PKEMRA. Congress may decide that some the requirements of PKEMRA have not been fulfilled or that certain requirements are no longer relevant.[82] Since the PPD-8 implementation process is in its early phases, it may be premature to assess comprehensive compliance with the statute. However, DHS is to submit a National Preparedness Report, due by March 30, 2012, intended to fulfill the reporting requirements in PKEMRA, notably 6 U.S.C. §752. The Report may be a valuable tool for Congress to assess the progress of the Executive Branch in fulfilling the requirements of PKEMRA. Additionally, Congress may wish to examine individual National Planning Frameworks, including the conformity of the National Disaster Recovery Framework to the PKEMRA mandate in 6 U.S.C. §77 1.[83]

Assignment of Federal Responsibilities

If, as predicted, a support function model is adopted by all of the National Planning Frameworks,[84] Congress may wish to evaluate the assignment of coordination and primary agency roles to federal departments and agencies. These roles typically carry a varied amount of additional responsibility for each Department assigned with a new role. Further, the Frameworks may attempt to coordinate and de-conflict the authorities and resources through this support function model, and may potentially do so in a manner that is disagreeable to Congress. Congress may also wish to review the leadership responsibilities dictated under PPD-8 policies. The President has directed the Secretary of the Department of Homeland Security (DHS), through the Assistant to the President for Homeland Security and Counterterrorism,[85] to lead the development of the National Preparedness Goal and System. HSPD-8 also directed the creation of the Goal and System by the Secretary of DHS. This delegation of leadership is an important distinction as it diverges from the language of PKEMRA that directed the President to develop the Goal and System by acting through the Administrator of FEMA .[86] Congress may wish to review the impact of this decision on the mission of FEMA and DHS and potential fulfillment of PKEMRA. Additionally, PPD-8 assigned leadership roles for the development and overall coordination of each National Planning Framework (see Table 2).

Effect on State and Local Governments, the Private Sector, and Citizens

PPD-8 policies are not, in principle, intended to place undue financial burden on state and local governments, the public and non-profit sector, and private citizens. Congress may want to evaluate PPD-8 guidance for its effect on these other government entities, especially for its compliance with the Unfunded Mandate Reform Act of 1995 (P.L. 104-4).[87] Further, Congress may wish to assess the effectiveness of the preparedness policies within their own districts and jurisdictions, and assess the overall preparedness level of their communities for the unique set of threats faced by each district. The capabilities identified in the National Preparedness Goal can serve as a guide for the types of tasks each state and local government will be expected to be able to accomplish, with the assistance of other stakeholders, in each mission area of prevention, protection, mitigation, response and recovery.

Impact on Federal Budget and Resources

In Paul Davis's definition of capability-based planning, he specifically noted that planning "is to be done not with the largesse of a blank-check policy (preparing for anything that might conceivably arise), but rather while working within an economic framework."[88] The National Preparedness Goal recognizes the need to use an economic framework in planning, as it notes:

> The results of these efforts and the specific deliverables called for in PPD-8 will inform current and future budget year planning and decisions. We will analyze current performance against our intended capabilities, the defined targets, and associated performance measures. This analysis will enable us to individually and collectively determine necessary resource levels, inform resource allocation plans, and guide Federal preparedness assistance. Budget implications across the preparedness enterprise cannot be assessed without this detailed and specific information. This approach will allow for annual adjustments based on updated priorities and our resource posture.[89]

Given current fiscal constraints, Congress may wish to assess, along with the Administration, the total budgetary implications of the "preparedness enterprise" developed by PDD-8. These implications may be specifically referenced in the President's future budget request for FY2012. For example, the establishment of multiple new National Planning Frameworks may bestow additional, and currently unfunded, responsibilities on federal agencies to prepare for and coordinate capabilities in the different mission areas. More holistically, if the standard for preparedness is based in part on the meta-scenario described in the draft National Preparedness Goal, Congress may wish to evaluate whether existing appropriated resources are sufficient to meet the challenge of the consequences described in the catastrophic scenario.

Monitoring and Evaluating the Overall Policy Impact of PPD-8

Finally, it is important to note that it may be difficult to assess the overall policy impact of PPD-8 in the near future. Because most of the evolution from HSPD-8 to PPD-8 is conceptual and strategic, the effect of PPD-8 policies on the preparedness of the nation for a disaster will take time to fully develop as concepts are operationalized. Further, some elements of PPD-8 policy may not be finalized until at least the end of September 2012. Though future exercises and trainings may provide an opportunity to evaluate some PPD-8 policies, a full analysis may not be possible until the nation either averts or is struck by a major disaster. This does not diminish the importance of PPD-8 policies, but may make it challenging for Congress to evaluate the relative worth of PPD-8 as a whole, and even more challenging to evaluate individual components of the policy.

End Notes

[1] The importance of national preparedness becomes starkly evident when evaluating the aftermath of the 2010 earthquake in Haiti and the 2011 Tōhoku earthquake and tsunami in Japan. Though both disasters were devastating, Japan's more advanced protective and mitigation measures, as well as their more robust capability to respond to and recover from the disaster, helped reduce the relative consequence of the disaster in Japan.

[2] P.L. 109-295. See 120 Stat. 1424-1432; 6 U.S.C. §741-754.

[3] White House, *Presidential Policy Directive 8: National Preparedness*, Washington, DC, March 30, 2011, http://www.dhs.gov/xabout/laws. Hereafter, the document will be referenced in footnotes as PPD-8.

[4] PPD-8, p. 1.

[5] Ibid.

[6] White House, *Homeland Security Policy Directive 8: National Preparedness*, Washington, DC, December 17, 2003. http://www.dhs.gov/files/publications/gc_1189788256647.shtm. Hereafter, the document will be referenced in footnotes as HSPD-8.

[7] Executive Order 12656, "Assignment of Emergency Preparedness Responsibilities," 53 *Federal Register* 47491, November 23, 1988.

[8] White House, *Presidential Policy Directive 1: Organization of the National Security Council*, Washington, DC, February 13, 2009, p. 5. Each Presidential administration has historically developed a new naming convention to apply to future Presidential directives. In PPD-1, the Obama Administration announced future directives would be named either a Presidential Policy Directive or Presidential Study Directive, replacing the old naming conventions of national security presidential directives and homeland security presidential directives.

[9] Memorandum from Randolph D. Moss, Acting Assistant Attorney General, to the Counsel of the President, on Legal Effectiveness of a Presidential Directive, As Compared to an Executive Order (Jan. 29, 2000), http://www.justice.gov/olc/predirective.htm.

[10] For more on Executive Orders, see CRS Report RS20846, *Executive Orders: Issuance and Revocation*, by Vanessa K. Burrows.

[11] However, some stipulations developed in policy directives have become requirements for federal grants, such as that states adopt and implement the National Incident Management System (NIMS). For more on this, see http://www.fema.gov/emergency/nims/CurrentYear Guidance.shtm#2010.

[12] For instance, though not obligated by law or regulation to do so, private sector owners and operators of critical infrastructure have participated willingly in the Sector Coordinating Councils of the National Infrastructure Protection Plan, outlined in White House, *Homeland Security Presidential Directive 8: Critical Infrastructure Identification, Prioritization, and Protection*, Washington, DC, December 17, 2003, http://www.dhs.gov/xabout/lawsgc_ 1214597989952.shtm. For more on Sector Coordinating Councils, see http://www.dhs. gov/files/partnerships/ editorial _0206.shtm.

[13] White House, *Implementation Plan for Presidential Policy Directive 8: National Preparedness*, Washington, DC, May 2011, p. 2. Hereafter document will referenced in footnotes as Implementation Plan for PPD-8.

[14] See White House, *National Security Strategy*, Washington, DC, May 2010, http://www. whitehouse.gov/sites/default/files/rss_viewer/national_security_strategy.pdf.

[15] Section 641(1) of PKEMRA, 6 U.S.C. §741.

[16] Implementation Plan for PPD-8, p. 2.

[17] For a description of this connection, see Sharon L. Claude, "Homeland Security Capabilities-Based Planning: Lessons from the Defense Community," *Homeland Security Affairs*, vol. I, no. 2 (Article 2, 2005).

[18] Paul K. Davis, *Analytic Architecture for Capabilities-Based Planning, Mission-System Analysis, and Transformation*, RAND National Defense Research Institute, Prepared for Office of the Secretary of Defense, Santa Monica, CA, 2002, pp. 1-2, http://www.rand.org/ pubs/monograph_reports/MR1513.

[19] Federal Emergency Management Agency, *Comprehensive Preparedness Guide 101: Developing and Maintaining Emergency Operations Plans* Version 2.0, Washington, DC, November 2010, p. B-2, http://www.fema.gov/pdf/about/ divisions/npd/CPG_101_V2.pdf.

[20] Ibid, p. 8.

[21] 42 U.S.C. §5195(a).

[22] 6 U.S.C. §742

[23] Though loosely defined, slow-onset disasters include threats like extended droughts, famines, gradual sea-level rise, and desertification (the expansion of arid or uninhabitable land).

[24] The National Mitigation Framework, discussed further on p. 17, may discuss some land-use and environmental policies, it is likely to do so in the context of mitigating the effects of immediate-onset natural disasters, as opposed to slow-onset.

[25] However, the National Response Framework may be revised to fit PPD-8 requirements. See "Content of the Frameworks."

[26] See Federal Emergency Management Agency, *National Preparedness Goal*, National Review Draft, Washington, DC, August 2011, http://www.fema.gov/prepared/ppd8.shtm. Hereinafter document will be referenced in footnotes as National Preparedness Goal, National Review Draft.

[27] Department of Homeland Security, *National Preparedness Goal*, First Edition, Washington, DC, September 2011, p. 1, http://www.fema.gov/pdf/prepared/npg.pdf. Hereinafter document will be referenced in footnotes as National Preparedness Goal.

[28] Under HSPD-8, the goal was:

> To achieve and sustain risk-based target levels of capability to prevent, protect against, respond to, and recover from major events, and to minimize their impact on lives, property, and the economy, through systematic and prioritized efforts by Federal, State, local and tribal entities, their private and non-governmental partners, and the general public.

[29] PPD-8's inclusion of mitigation arguably conforms more closely with the intention of PKEMRA. See 6 U.S.C. §743(a).

[30] PP. 6 of PPD-8.

[31] See National Preparedness Goal, p. 3, footnote 4. The "Latest Updates" section of the FEMA PPD-8 website indicates this document will be online soon, see http://www.fema.gov/prepared/ppd8.shtm.

[32] National Preparedness Goal, p. 7.

[33] 6 U.S.C. §744(c) and 745.

[34] See http://www.fema.gov/pdf/media/factsheets/2009/npd_natl_plan_scenario.pdf.

[35] On p. 6 of the National Preparedness Goal, National Review Draft, the meta-scenario is described as:

> There is a no-notice event impacting a population of seven million within a 25 thousand square mile area. The impacted area includes several states across multiple regions. Severe damage is projected to critical infrastructure including essential transportation infrastructure. Ingress and egress options are severely limited. The projected number of fatalities is 195,000 during the initial hours of the event. It is projected that 265,000 survivors will require emergency medical attention. At least 25 percent of the impacted population will require mass care, emergency sheltering, and housing assistance.

[36] National Preparedness Goal, p. 4.

[37] The methodology used by DHS for the analysis has a considerable impact on the specificity and types of results that could be produced in the assessment. For a full discussion on risk analysis in DHS, see Committee to Review the Department of Homeland Security's Approach to Risk Analysis; National Research Council, *Review of the Department of Homeland Security's Approach to Risk Analysis* (Washington, DC: The National Academies Press, 2010), http://www.nap.edu/catalog.php?record_id=12972.

[38] See National Preparedness Goal, p. 3, footnote 6. A more comprehensive description of the Strategic National Risk Assessment may be available online soon, according to the "Latest Updates" section of the FEMA website on PPD-8. See http://www.fema.gov/prepared/ppd8.shtm.

[39] National Preparedness Goal, p. 4.

[40] PPD-8, p. 3.

[41] An IPC is described on p. 5 of Presidential Policy Directive 1 as "the main day-to-day fora for interagency coordination of national security policy. They shall provide policy analysis for consideration by the more senior committees of the NSC system and ensure timely responses to decisions made by the President." Specific descriptions of the purpose of each IPC were not provided by the White House. See White House, *Presidential Policy Directive 1: Organization of the National Security Council*, Washington, DC, February 13, 2009.

IPCs replace "Policy Coordination Committees" of the George W. Bush Administration. For more on the purpose and history of the National Security Council, see CRS Report RL30840, *The National Security Council: An Organizational Assessment*, by Richard A. Best Jr.

[42] National Preparedness Goal, p. 4-18.

[43] The Prevention Framework is to include capabilities needed to conduct investigations following a terrorist attack during the response phase to a disaster (Implementation Plan, p. 9).

[44] They are "Planning," "Public Information and Warning," and "Operational Coordination."

[45] For instance, the National Prevention Framework will probably leverage authorities stemming from legislation including, but not limited to, the Intelligence Reform and Terrorism Prevention Act (P.L. 108-458), the Implementing Recommendations of the 9/11 Commission Act of 2007 (P.L. 110-53), and establishing legislation for the national intelligence and law enforcement agencies. The National Response Framework lists 15 principle emergency authorities, 48 other statutory authorities and regulations, 17 executive orders, and 20 presidential directives as guiding the structure, development, and implementation of the NRF. For a full list, see U.S. Department of Homeland Security, Federal Emergency Management Agency, *National Response Framework: List of Authorities and References*, January 2008, pp. 1-16, http://www.fema.gov/pdf/emergency/nrf/nrf-authorities.pdf.

[46] PPD-8, p.3.

[47] Because this section is predictive in nature, it cannot be absolutely authoritative. The suppositions and predictive analysis are based on the expertise of CRS analysts.

[48] *Implementation Plan*, p. 4.

[49] Department of Homeland Security, *National Disaster Recovery Framework*, Washington, DC, September 2011, http://www.fema.gov/pdf/recoveryFramework/ndrf.pdf.

[50] Department of Homeland Security, *National Response Framework*, Washington, DC, January 2008, http://www.fema.gov/pdf/emergency/nrf/nrf-core.pdf.

[51] Department of Homeland Security, *National Infrastructure Protection Plan: Partnering to Enhance Protection and Resiliency*, Washington, DC, 2009, http://www.dhs.gov/xlibrary/assets/NIPP_Plan.pdf.

[52] PPD-8, p. 3.

[53] The Sector, or Sector-Specific Agency model, of the NIPP aligns responsibilities by physical or subject area of expertise, as opposed to capabilities. For instance, the Environmental Protection Agency is assigned as the Sector Specific Agency for the water sector because of their leading expertise on all matters relating to water infrastructure, regardless of capability. For more on this approach, see CRS Report RL30153, *Critical Infrastructures: Background, Policy, and Implementation*, by John D. Moteff.

[54] PPD-8, p. 3.

[55] For a description of how this applies in the NRF, see Federal Emergency Management Agency, *Emergency Support Function Annexes: Introduction*, Washington, DC, January 2008, http://www.fema.gov/pdf/emergency/nrf/nrf-esfintro.pdf.

[56] For more on this topic, see CRS Report R40901, *Terrorism Information Sharing and the Nationwide Suspicious Activity Report Initiative: Background and Issues for Congress*, by Jerome P. Bjelopera.

[57] For more on this issue, see White House, *Empowering Local Partners to Prevent Violent Extremism in the United States*, Washington, DC, August 2011.

[58] An abbreviated list of federal agencies that may be included in the Framework are the: Bureau of Alcohol, Tobacco, Firearms and Explosives; Central Intelligence Agency; Customs and Border Patrol; Department of Defense intelligence agencies; Drug Enforcement Agency; Federal Bureau of Investigation; National Security Agency; National Protection and Programs Directorate, Office of Intelligence and Analysis; Transportation Security Administration; and the U.S. Coast Guard.

[59] A few examples include

- On continuity of operations; Federal Emergency Management Agency, *Federal Continuity Directive 1: Federal Executive Branch National Continuity Program and Requirements*, February 2008, http://www.fema.gov/pdf/about/offices/fcd1.pdf.
- On supply chain security; Department of Homeland Security, *Strategy for International Supply Chain Security* , Washington, DC, July 2007, http://www.dhs.gov/xlibrary/assets/plcyinternationalsupplychainsecuritystrategy.pdf. This document is currently being revised.c
- On cyber security; the White House led "Comprehensive National Cybersecurity Initiative,"see http://www.whitehouse.gov/cybersecuritycomprehensive-national-cyber security-initiative.

[60] PPD-8, p. 6.

[61] National Preparedness Goal, p. 9.

[62] For more on FEMA's programs in hazard mitigation, see CRS Report RL34537, *FEMA 's Pre-Disaster Mitigation Program: Overview and Issues*, by Francis X. McCarthy and Natalie Keegan and CRS Report R40471, *FEMA 's Hazard Mitigation Grant Program: Overview and Issues*, by Natalie Keegan.

[63] See CRS Report RL34537, *FEMA 's Pre-Disaster Mitigation Program: Overview and Issues*, by Francis X. McCarthy and Natalie Keegan.

[64] For more on this, see CRS Report R41752, *Locally Operated Levees: Issues and Federal Programs* , by Natalie Keegan et al.

[65] Department of Homeland Security, *National Response Framework*, Washington, DC, January 2008, http://www.fema.gov/pdf/emergency/nrf/nrf-core.pdf.

[66] For more on the development of the NRF, see CRS Report RL34758, *The National Response Framework: Overview and Possible Issues for Congress*, by Bruce R. Lindsay, *The National Response Framework: Overview and Possible Issues for Congress*, by Bruce R. Lindsay.

[67] Guidance may especially be needed on how the NRF is changed by the elimination of Emergency Support Function #14: Long-Term Community Recovery by the NDRF, and how the National Prevention Framework capability "Forensics and Attribution" is incorporated within overall disaster response.

[68] On p. 6 of the National Preparedness Goal, National Review Draft, the meta-scenario is described as:

> There is a no-notice event impacting a population of seven million within a 25 thousand square mile area. The impacted area includes several states across multiple regions. Severe damage is projected to critical infrastructure including essential transportation infrastructure. Ingress and egress options are severely limited. The projected number of fatalities is 195,000 during the initial hours of the event. It is projected that 265,000 survivors will require emergency medical attention. At least 25 percent of the impacted population will require mass care, emergency sheltering, and housing assistance.

[69] Federal Emergency Management Agency, *National Response Framework: Catastrophic Incident Annex*, Washington, DC, November 2008, http://www.fema.gov/pdf/emergency/nrf/nrf_CatastrophicIncidentAnnex.pdf.

[70] Department of Homeland Security, *National Disaster Recovery Framework*, Washington, DC, September 2011, http://www.fema.gov/pdf/recoveryFramework/ndrf.pdf

[71] Section 682 of PKEMRA (6 U.S.C. §771) called for a "National Disaster Recovery Strategy" by July 1, 2007. FEMA has referenced the National Disaster Recovery Framework as fulfilling this mandate.

[72] See http://www.disasterworkingrecoveryworkinggroup.gov.

[73] According to a schedule in the "Purpose Statement" section on the Working Group website, the *Report to the President* was due to be completed by first of April, 2010.

[74] The six RSFs are: Community Planning and Capacity Building; coordinated by FEMA; Economic, coordinated by Department of Commerce; Health and Social Services, coordinated by HHS; Housing, coordinated by HUD; Infrastructure Systems; coordinated by USACE; and Cultural and Natural Resources, coordinated by Department of Interior.

[75] See NDRF, p. 25.

[76] Department of Homeland Security, "National Disaster Recovery Framework," 75 *Federal Register* 6681, February 10, 2010.

[77] The Committee submitted a formal letter to the Secretary of DHS requesting it be included as part of the public record of the Federal Register notice. The Document ID for this public submission is: FEMA-2010-0004-0127. See http://www.regulations.gov/#!documentDetail; D=FEMA-2010-0004-0127

[78] 6 U.S.C. §771(b)(1)

[79] 6 U.S.C. §771(b)(4)

[80] For more on the PS-Prep program, see http://www.fema.gov/privatesector/preparedness/.

[81] See p. 11 of the Implementation Plan.

[82] For instance, the National Planning Scenarios requirement (6 U.S.C. §744(c) and 745).

[83] See "Potential Issues and Challenges" on the NDRF.

[84] See "Coordinating Structure and Support Function Model."

[85] This is a position in the National Security Council, currently held by John Brennan.

[86] 6 U.S.C. §743-744. It is notable, however, that the main website for PPD-8 is hosted on the FEMA website, at http://www.fema.gov/prepared/ppd8.shtm.

[87] For more on the Unfunded Mandate Reform Act, see CRS Report R40957, *Unfunded Mandates Reform Act: History, Impact, and Issues*, by Robert Jay Dilger and Richard S. Beth.

[88] Paul K. Davis, *Analytic Architecture for Capabilities-Based Planning, Mission-System Analysis, and Transformation*, RAND National Defense Research Institute, Prepared for Office of the Secretary of Defense, Santa Monica, CA, 2002, pp. 1-2, http://www.rand.org/ pubs /monograph_reports/MR1513.

[89] National Preparedness Goal, p. 19.

In: Federal Emergency Management
Editor: Lewis B. Adams
ISBN: 978-1-62618-217-2

Chapter 3

FEDERAL STAFFORD ACT DISASTER ASSISTANCE: PRESIDENTIAL DECLARATIONS, ELIGIBLE ACTIVITIES, AND FUNDING*

Francis X. McCarthy

SUMMARY

The Robert T. Stafford Disaster Relief and Emergency Assistance Act (the Stafford Act) authorizes the President to issue major disaster or emergency declarations in response to catastrophes in the United States that overwhelm state and local governments. Such declarations result in the distribution of a wide range of federal aid to individuals and families, certain nonprofit organizations, and public agencies. Congress appropriates money to the Disaster Relief Fund (DRF), through both annual appropriations and emergency supplemental appropriations, for disaster assistance authorized by the Stafford Act. The Federal Emergency Management Agency (FEMA) within the Department of Homeland Security (DHS) administers most, but not all, of the authority the statute vests in the President.

The most recent significant action concerning the statute occurred in the closing months of the 109th Congress as a result of the congressional

---

* This is an edited, reformatted and augmented version of a Congressional Research Service publication, CRS Report for Congress RL33053, from www.crs.gov, dated June 7, 2011.

investigation on the response to Hurricane Katrina (August 2005). Senators inserted Stafford Act amendments into the FY2007 DHS appropriations legislation (Title VI of P.L. 109-295).

These amendments expanded FEMA's authority to expedite emergency assistance to stricken areas, imposed new planning and preparedness requirements on federal administrators, provided new authority to regional offices, and increased federal assistance to victims and communities.

More recently, Congress included a provision in the FY2010 appropriations legislation (P.L. 111-83) that allows retired law judges to arbitrate conflicts concerning the recovery of public infrastructure in the Gulf Coast due to Hurricanes Katrina and Rita. While not an amendment to the Stafford Act, this provision affects the administration of the FEMA appeals process under which applications for Stafford assistance are reconsidered. The decisions made to date by the arbitration panels resulted in an Administration request for supplemental funding that resulted in P.L. 111-112, which added more than $5.5 billion to the Disaster Relief Fund.

Previously introduced legislation during the 111th Congress would have amended the statute. Among the proposals, H.R. 3377, the Disaster Response, Recovery, and Mitigation Enhancement Act of 2009, would have authorized the Disaster Relief Fund, provided health benefits to temporary or intermittent federal employees who provide disaster assistance, authorized the National Urban Search and Rescue Response System, and requested FEMA to update standards for individual assistance disaster requests.

Other bills sought to reauthorize a mortgage and rental assistance program terminated in 2000 (H.R. 888/S. 763), establish new eligibility criteria (H.R. 941, H.R. 1059, H.R. 1494, H.R. 2484, H.R. 4141, and S. 1069), and mandate establishment of a tracking and storage plan for housing units used by disaster survivors (H.R. 3437/S. 713).

OVERVIEW OF THE STAFFORD ACT

The Robert T. Stafford Disaster Relief and Emergency Assistance Act (the Stafford Act) authorizes the President to issue major disaster, emergency, and fire management declarations, which in turn enable federal agencies to provide assistance to state and local governments overwhelmed by catastrophes.[1]

The statute also authorizes the President to determine whether certain types of authorized assistance will be provided and the conditions under which the aid is distributed.

The Stafford Act comprises the following titles:

- Title I: findings, declarations, and definitions;
- Title II: preparedness and pre-disaster mitigation authority;
- Title III: administrative provisions;
- Title IV: major disaster assistance;
- Title V: emergency assistance;
- Title VI: emergency preparedness (previously the Civil Defense Act of 1950);
- Title VII: miscellaneous provisions.

Titles IV and V identify the types of assistance that may be provided, and, in some circumstances, the limitations on the aid.[2] Most of the presidential authority set out in Titles IV and V in the statute, with the exception of the authority to issue declarations, has been delegated to administration officials—currently the Secretary of the Department of Homeland Security (DHS)—through executive orders.[3] Following the investigation into the response to Hurricane Katrina, Congress also authorized the Administrator to provide federal leadership before and after catastrophes, including "assisting the President in carrying out the functions" of the Stafford Act.[4]

Stafford Act definitions establish eligibility for assistance and the conditions that may be catalysts for presidential action. The definition of "major disaster" is relatively restrictive – "any natural catastrophe" including storms, earth movements and high water, and "regardless of cause, any fire, flood or explosion."[5] Presidents Bush and Obama issued 55 major disaster declarations in calendar year 2009.[6] By comparison, the definition of "emergency" is broad, authorizing the President to determine "any occasion or instance" when federal aid is needed by state and local governments to save lives and property or to address the threat of a catastrophe.

In calendar year 2009, President Bush issued one emergency declaration for the presidential inauguration in the District of Columbia; President Obama issued six others in the remainder of the year, five for winter storms and one for a refinery explosion in Puerto Rico.[7] The statute also authorizes the President to provide fire suppression assistance to prevent a forest or grassland fire from becoming a major disaster.[8] President Obama issued 48 fire management declarations in 2009; President Bush issued one before leaving office, for a yearly total of 49 such declarations.[9]

Stafford Act assistance funding derives from appropriations made to the Disaster Relief Fund (DRF), administered by DHS. [10] Federal assistance supported by the DRF, and authorized in Title IV (for major disasters) or Title V (for emergencies) provides grants for mass care for disaster survivors, the restoration of damaged or destroyed facilities, amelioration of the impact of future disasters, clearance of debris, and aid for those with uninsured critical needs. The statute also authorizes loans to communities that suffer significant revenue losses as a result of major disasters.[11] In addition, the statute authorizes unemployment assistance directly related to the event (administered by the Department of Labor) and allows federal agency heads to provide technical assistance, personnel, equipment and other resources to help state and local response and recovery efforts.

While the President may exercise discretion under authority of the statute, that authority is circumscribed by the act's administrative provisions. Federal aid cannot duplicate that provided by other sources. Federal agency heads may waive administrative, but not statutory, requirements to expedite assistance. A federal coordinating officer, appointed by the President (or the designee), ensures that federal and non-federal assistance provided to major disaster or emergency areas is coordinated. Applicants for aid must purchase appropriate levels of insurance. Title III of the statute sets out these and other administrative requirements and limitations.

The 1974 Disaster Relief Act, as Amended

The Stafford Act consists of the text of the statute enacted by Congress in 1974 and that of a series of amendments approved over the past 35 years that have added administrative requirements, enhanced the types of assistance that may be provided, expanded eligibility to certain non-profit organizations, and coordinated Stafford assistance with other federal authorities.[12]

The act maintains a two-part orientation first established in a 1950 statute (superseded by the 1974 legislation), as follows: (1) federal assistance supplements state and local relief and recovery efforts and, (2) is triggered only by a presidential declaration that is preceded by a gubernatorial request for assistance.[13]

The scope of the statute has expanded considerably over the decades to provide a wide range of grants for the needs of individuals, families, certain community organizations, and state and local government operations. The congressional investigation initiated to address questions about the response to

Hurricane Katrina, which struck in August 2005, culminated in enactment of amendments that established new levels of assistance, but perhaps more importantly, addressed administrative concerns. The following section highlights the more significant amendments adopted in 2006.

The 2006 Amendments

Through the Post-Katrina Emergency Management Reform Act of 2006 (Title VI, P.L. 109-295) and, to a lesser extent, the Security and Accountability for Every Port Act of 2005, known as the SAFE Port Act (P.L. 109-347), Congress expanded the authority of FEMA to provide assistance and imposed requirements on federal officials to ensure that effective pre-disaster preparation actions would be taken.[14] Assistance for service animals and pets, pre-disaster authority, evacuation planning, and individual case management are some of the enhancements included in these amendments. While augmenting federal assistance authorities such as these, the amendments maintained state, local, and individual emergency management responsibility and accountability. In short, the Post-Katrina Act expanded the President's federal disaster assistance authority, but left the basic tenets of the Stafford Act (such as Presidential discretion, need for state requests, most restrictions on eligibility) unchanged.

The changes to the Stafford Act enacted by Congress in 2006 may be summarized as follows:

- Expedited federal assistance. The President is now authorized to support precautionary evacuation measures, accelerate federal emergency response and recovery aid, and provide expedited federal assistance (coordinated with the state to the extent possible) in the absence of a specific request from state officials.[15]
- Aid to individuals with special needs. Disabled individuals now receive federal housing assistance on sites that meet their needs, "whenever practicable," with the inclusion in the Stafford Act of the definition of "individual with a disability" from the Americans With Disabilities Act of 1990.16 Also, durable medical equipment, such as that needed by the disabled, may be provided,17 and the FEMA Administrator must develop guidelines concerning the accommodation of individuals with disabilities with regard to emergency facilities and equipment.[18] The amendments also prohibit discrimination toward the disabled, those with special needs, and those with limited proficiency in the English language.[19] To assist the

latter segment of disaster victims, the FEMA Administrator must ensure that information is made available to those with limited English proficiency, and must develop and maintain information on model language assistance as well as best practices for the state and local governments working with these individuals.[20] Also, the Pets Evacuation and Transportation Standards Act of 2006 authorizes grants to states for the improvement of emergency shelters to accommodate pets and service animals.[21]

- *Expanded assistance to disaster victims.* The President is authorized to provide transportation assistance to those displaced from their residences, including aid needed to move among alternative temporary shelters or to return to their original residence. The President may also provide case management services to state, local, or qualified private organizations that provide assistance to victims.[22]
- *Housing assistance.* Seven provisions in the 2006 amendments expanded the scope of federal housing assistance to be made available to victims.[23] First, in order to be considered eligible for housing assistance, victims of major disasters or emergencies who are disabled must be unable to access or inhabit their homes.[24] Prior to enactment of the 2006 amendments the statute did not differentiate disabled access from that experienced by other disaster victims. Second, alternative housing sites provided to victims by the federal government must meet physical accessibility requirements.[25] Third, the amendments eliminated the statutory ceilings on financial aid to be provided for housing repair and replacement,[26] but did not eliminate the overall cap of $25,000 that may be provided to each individual or household under Section 408.[27] Fourth, the amendments authorize the President to provide semi-permanent housing where no alternatives exist.[28] Fifth, the statute includes as newly eligible housing-assistance costs both utility costs (excluding telephone service) and security deposits.[29] Sixth, the disposal of temporary housing units (generally referred to as "FEMA trailers") is to be coordinated with the Department of the Interior or other federal agencies to facilitate the transfer of the units to tribal governments.[30] Seventh, the amendments established a demonstration project, the Individuals and Households Pilot Program, to increase the use of existing rental housing to provide temporary housing for victims of major disasters. Through the pilot program, which expired December 31, 2008, the Administrator was to provide for the repair and

improvement of multi-family rental properties in disaster areas to increase the rental stock available to disaster victims in the immediate area.[31]

- *Public assistance (PA) to state and local governments.* State or local governments or eligible private nonprofit-facility owners may apply for federal assistance to build a new facility in a different location if all parties determine that a damaged facility should not be repaired or replaced. Such assistance, known as the "in-lieu" contribution (a grant made in-lieu of replacing or repairing a facility that existed before the disaster), has been authorized since 1974 and consists of PA grants based on the federal estimate of the cost of repair or replacement.[32] The 2006 amendment in the SAFE Ports Act deleted the clause that authorized an in-lieu grant for 90% of the federal share of the estimate of repairing or replacing the facility solely in areas with unstable soil. Now all stricken state and local governments, not just those in areas marked by unstable soil, may apply for a 90% in-lieu grant.[33] The SAFE Ports Act also authorized expedited payments for debris removal to state or local governments or to owners of qualified private nonprofit facilities.[34] In addition, the SAFE Ports Act amended the Community Disaster Loan (CDL) provisions by authorizing loans to local governments (capped at $5 million) that, because of a major disaster, suffer significant losses in tax revenue.[35] Another PA change required the President, through the FEMA Administrator, to conduct a pilot program of incentives for local and state government involvement in debris removal and the acceleration of repair work.[36]
- *Assistance to nonprofit organizations.* Three significant changes were adopted regarding the federal aid provided to private nonprofit organizations affected by a major disaster. First, the statute authorizes the President, within stated constraints, to define facilities that provide "essential services of a governmental nature to the general public."[37] These provisions, similar to those in FEMA regulations, establish eligibility for museums, zoos, performing arts facilities, community arts centers, and other facilities that "provide health and safety services of a governmental nature" as long as they provide "essential services of a governmental nature to the general public," as defined by the President.[38] Third, the amendments added the word "education" to the listing in the section of the law that defines critical services. Through this change, private nonprofit education organizations may

apply directly for a FEMA grant without having to apply for a Small Business Administration loan.[39]

- *Hazard mitigation.* Section 404 of the Stafford Act authorizes the Hazard Mitigation Grant Program (HMGP) for grants to states in which major disasters have been declared.[40] These funds must be used for activities that prevent future disasters or reduce their impact if they cannot be prevented. The Post-Katrina Act adjusted the percentage amounts for HMGP awards by establishing a scale that authorizes a higher percentage (15% of the total Stafford Act assistance in a state) for major disasters in which no more than $2 billion is provided, 10% for assistance that ranges from more than $2 billion to $10 billion, and 7.5% for major disasters that involve Stafford Act assistance of more than $10 billion to $35.3 billion.[41]
- *Administrative changes.* The 2006 amendments authorize the President to appoint one Federal Coordinating Officer (FCO) for a multi-state event, in addition to deputy FCO's as needed.[42] Traditionally, one FCO has been named for each separate disaster declaration in each respective state. The statute also amends the Stafford Act by requiring that the President designate a Small State and Rural Advocate in FEMA to ensure that rural community needs are met in the declaration process and to help small states prepare declaration requests, among other duties.[43] Other administrative changes concern the authority and capabilities of organizations charged with the response to the major disaster site. The amendments authorize the President to establish at least three national response teams and others as deemed necessary (including regional response teams), and require that FEMA team members possess essential capabilities, training skills, and equipment.[44] In addition, the SAFE Ports Act amended the Stafford Act by adding a new definition, "essential service providers," defined as persons affiliated with municipal governments or private entities who will help restore essential services to a stricken area. Such workers are not to be impeded when they seek access to a disaster site.[45]

Action in the 111th Congress

Legislation introduced in the 111th Congress could have expanded categories of federal assistance in the Stafford Act and, if enacted, make relatively minor adjustments to the statute. Selected examples of the legislation and summaries of the policy changes that would be incorporated into the statute are listed below:

- *H.R. 888 and S. 763:* Temporary mortgage and rental assistance would be authorized for victims in areas included in a major disaster declaration. Congress previously authorized the assistance from 1974 until enactment of the 2000 amendments which vitiated the provision as part of the efforts by the 106th Congress and the Clinton Administration to constrain federal emergency management costs.
- *H.R. 941 and S. 1069:* H.R. 941 would designate owners or operators of a private or investor-owned electric utility company serving low-income households to be eligible for grants for the repair or replacement of damaged facilities. S. 1069 would do the same for owners or operators of investor owned energy facilities, under certain conditions. As explained later in this report (see "Expanding Eligibility" section) this issue has been debated in the past as certain utilities that provide public services have no recourse to Stafford assistance.
- *H.R. 1239:* States could apply for grants to establish revolving loan funds to enable homeowners and businesses to undertake hazard mitigation activities. This legislation would amend the pre-disaster mitigation authority (Section 203 of the act). Experience gained under the mitigation efforts funded through a pilot program established through the Bunning-Bereuter-Blumenauer Flood Insurance Reform Act of 2004 for severe repetitive loss properties in flood plains may be applicable to congressional consideration of this proposal.[46]
- *S. 713 and H.R. 3437*: These two bills would modify existing disaster housing provisions. If enacted, H.R. 3437 would mandate that the FEMA Administrator develop lifecycle plans and tracking procedures for mobile housing units provided to individuals and households. Units found to be hazardous to the occupants' health would be repaired before being used for disaster-related housing needs.[47] Similarly, but not as an amendment to the Stafford Act, S. 713 would mandate that the FEMA Administrator evaluate the existing stock of temporary housing units and plan for their disposition or storage.
- *H.R. 3466:* The President would be authorized to consolidate funds for public infrastructure restoration for disasters involving "extensive and widespread damage and destruction" with no cost share required from state or local governments.[48]

- *S. 764:* The ceiling on the amount to be available for individual and household assistance for lower income residents in states with high costs of living would be increased to a base of $50,000, from $25,000, to be adjusted annual for inflation.
- *H.R. 3377*: A number of changes would be made to the Stafford Act through this legislation. Funding would be authorized through FY2012 for the Pre-Disaster Mitigation program (described below in this report), the Administrator of FEMA would be required to modernize and modify the existing federal public warning and alert system, the Mortgage and Rental Assistance program would be reauthorized, temporary or intermittent employees who provide disaster assistance would be eligible for federal health benefits, the National Urban Search and Rescue Response system would be authorized, the Disaster Relief Fund would be authorized, and grants for hazard mitigation would be increased if states meet minimum building code standards.[49]

Enacted Legislation

While no amendments to the Stafford Act were enacted in the 111th Congress, two public laws should be noted for the measures they include that affect administration of the statute. First, provisions in a supplemental appropriations statute for FY2009 authorized the FEMA Administrator to extend the payment for case management programs related to Hurricanes Katrina and Rita and mandated a 90/10 cost share for public assistance stemming from Hurricane Ike, storms in Kentucky and Kansas, among other purposes.[50] Second, the FY2010 appropriations statute for DHS authorizes the reemployment of retired administrative law judges to settle disputes related to the arbitration panels established in the American Recovery and Reinvestment Act of 2009 in order to expedite recovery from Hurricanes Katrina and Rita. This change in procedure is discussed at the end of this report, see "Appeal Process" section, below.[51]

Overview of Stafford Act Declarations

Major disaster and emergency declarations are two of the five types of actions that may be taken under authority of the Stafford Act.[52] The other three include fire management declarations, the provision of defense resources before a major disaster is declared, and the decision to pre-position supplies and resources.[53]

Prior to a Disaster

Three types of declarations (or commitments) may be made under Stafford Act authority before a catastrophe occurs. First, at the request of a governor, the President may direct the Department of Defense (DOD) to commit resources for emergency work essential to preserve life and property in "the immediate aftermath of an incident" that may result in the declaration of a major disaster or emergency (discussed below).[54] The statute does not define the term "incident." According to regulations, upon receiving a gubernatorial request for such assistance, the FEMA Associate Director may determine that DOD aid is necessary to save lives and protect property.[55]

Second, the Stafford Act authorizes the President to provide fire management assistance in the form of grants, equipment, personnel, and supplies to supplement the resources of communities when fires on public property or on private forests or grasslands threaten destruction that might warrant a major disaster declaration.[56] Implementation of this authority, which has been delegated to FEMA officials, requires that a gubernatorial request be submitted while an uncontrolled fire is burning.

To be approved, state applications must demonstrate that either of the two cost thresholds established by FEMA through regulations has been reached.[57] The thresholds involve calculations of the cost of an individual fire or those associated with all of the fires (declared and non-declared) in a state each calendar year.[58] FEMA officials determine whether a fire management assistance declaration will be issued.[59]

Third, when a situation threatens human health and safety, and a disaster is imminent but not yet declared, the Secretary of DHS may pre-position employees and supplies.

DHS monitors the status of the situation, communicates with state emergency officials on potential assistance requirements, deploys teams and resources to maximize the speed and effectiveness of the anticipated federal response and, when necessary, performs preparedness and preliminary damage assessment activities. A related provision enacted in 2006 authorizes the FEMA Administrator to provide evacuation preparedness assistance to non-federal officials.[60]

After a Catastrophe Occurs

The Stafford Act authorizes the President to issue two types of declarations—major disaster and emergency—after an incident overwhelms state and local resources.[61]

Major Disaster Declarations

After receiving a request from the governor of an affected state for a major disaster declaration, the President may take one of three possible actions: issue a major disaster declaration or an emergency declaration, or decline the request.[62] Major disaster declarations may be issued after a natural catastrophe "(including any hurricane, tornado, storm, high water, wind-driven water, tidal wave, tsunami, earthquake, volcanic eruption, landslide, mudslide, snowstorm, or drought)" or, "regardless of cause, [after a] fire, flood or explosion."[63]

Regulations further specify the factors considered by FEMA in evaluating a gubernatorial request for a major disaster declaration. The factors considered to determine whether federal PA assistance is needed include an assessment of the per capita impact of the disaster within affected states;[64] insurance coverage in force; the presence and impact of hazard mitigation measures; the cumulative impact of disasters over the previous year; and, whether federal aid authorized by statutes other than the Stafford Act would better meet the needs of stricken areas.[65]

Factors considered to determine whether federal IHP assistance is needed include concentration of damages; number of injuries, deaths, or the extent to which essential services are disrupted; the impact on special populations that require higher levels of assistance; the extent to which voluntary agencies are able to meet the needs of victims; insurance coverage; and, measurements of needs such as disaster housing needs approved, number of homes destroyed or damaged, financial assistance required, and others.[66]

Emergency Declarations

The declaration process for emergencies is similar to that used for major disasters, but the criteria (based on the definition of "emergency") are less specific.[67] Whereas all major disaster declarations require a gubernatorial request and, generally, findings and certifications as summarized above, emergency declaration requirements are less rigorous. For example, the President may issue an emergency declaration without a gubernatorial request if primary responsibility rests with the federal government.[68] Also, specific thresholds or calculations of past averages are not considered, but FEMA officials do assess whether "all other resources and authorities available to meet the crisis are inadequate" before recommending that the President issue an emergency declaration.[69]

TYPES OF ASSISTANCE AND ELIGIBILITY

The Stafford Act identifies the universe of eligible applicants (e.g., states, local governments, owners of certain private nonprofit facilities, individuals, or families). However, not all persons or entities affected by a catastrophe are eligible for Stafford Act assistance even if the President issues a declaration. FEMA officials determine the need for assistance after a major disaster or emergency declaration is issued.

Aid is provided only to those persons or entities determined to need the assistance. For example, a family with adequate insurance and alternative housing options might not be considered eligible to receive financial aid. A unit of local government that suffers damages to some facilities, but not to the extent considered necessary pursuant to FEMA regulations and guidelines, might not receive funds to rebuild infrastructure.

The type of assistance made available varies from one disaster to another and among eligible applicants within a state, commensurate with decisions by FEMA officials on the extent of damage and the eligibility of applicants. For instance, under a major disaster declaration, local jurisdictions with large numbers of damaged or destroyed residences might be eligible for assistance under the Individual Assistance (IA) program, whereas those with severely damaged infrastructure but relatively few damaged homes might be eligible only for assistance under the PA program.

Similarly, if a local government had extensive debris in public rights-of-way due to a disaster, but very little damage to public facilities, a determination might be made to provide assistance only for debris removal activities under the PA program. On the other hand, areas severely devastated by a catastrophe are often eligible for both IA and PA.

The Stafford Act authorizes the President to make the initial determination of eligibility for federal relief and recovery assistance through the issuance of either a major disaster or emergency declaration. The following subsections summarize the types of assistance authorized under each.

Major Disaster Assistance

A major disaster declaration authorizes the President to direct that the following types of federal disaster assistance may be provided, depending upon the specific needs of the stricken areas:

- *General federal assistance* supports state and local governments in that it facilitates the distribution of consumable supplies; authorizes federal agency heads to provide resources (with or without reimbursement) to support evacuations, response and recovery efforts; and, provides a range of technical and advisory help.[70]
- *Essential assistance* authorizes federal agency heads to distribute aid to victims through state and local governments and voluntary organizations, perform life- and property-saving assistance, clear debris, and conduct search and rescue missions, among other immediate response services. (The statute also allows Governors to request the use of Department of Defense resources before a major disaster or emergency declaration is issued.)[71]
- *Hazard mitigation* grants are provided to each state in which a major disaster declaration is issued to reduce risks and damages that might occur in future disasters.[72]
- *Federal facilities* may be repaired, restored, replaced, or reconstructed, if necessary, before legislation is enacted. If necessary, such construction may be undertaken before appropriations are approved if funds can be transferred from accounts dedicated to other purposes.[73]
- *Repair, restoration, and replacement of damaged facilities* is administered through the Public Assistance (PA) program. Public facilities and infrastructure systems owned by state and local governments, as well as private nonprofit facilities that provide essential services, may be repaired or rebuilt. Applicants may also seek contributions for other facilities or hazard mitigation measures in lieu of repairing or restoring damaged facilities;[74]
- *Debris removal* may be funded through the use of federal resources or through grants to state or local governments or owners of private nonprofit facilities.[75]
- *Assistance to individuals and households* administered through the Individual and Household Program (IHP), includes financial grants to

rent alternative housing, direct assistance through temporary housing units (mobile homes), and limited financial assistance for housing repairs and replacement. Financial assistance for uninsured medical, dental, funeral, personal property, transportation, and other expenses (referred to as Other Needs Assistance (ONA) is also authorized.[76]

- *Unemployment assistance* is available to individuals unemployed as a result of the major disaster, for up to 26 weeks, as long as they are not entitled to other unemployment compensation or credits.[77]
- *Grants to assist low-income migrant and seasonal farmworkers* are provided by the Secretary of Agriculture (total limited to $20 million annually) "where the Secretary determines that a local, state or national emergency or disaster" has resulted in a loss of income or inability to work.[78]
- *Food coupons and food distribution* for low-income households unable to purchase nutritious food;[79]
- food commodities for emergency mass feeding;[80]
- legal services for low-income individuals;[81]
- crisis counseling assistance and training grants for state and local governments or private mental health organizations to provide services or train workers;[82]
- community disaster loans to local governments that lose tax or other revenues needed for governmental services;[83]
- emergency communications to establish temporary communications during "or in anticipation of an emergency or major disaster,";[84] and,
- emergency public transportation to provide transportation to essential places.[85]

Each major disaster declaration specifies the type of incident covered, the time period covered, the types of disaster assistance available, the units of local government (generally counties, parishes, and independent cities) included in the declaration, and the name of the federal coordinating officer. As the effects of the catastrophe subside over time, the initial major disaster declaration may be amended to modify the types of assistance to be provided and the areas (generally counties) included in the major disaster declarations.

Emergency Declaration Assistance

Considerably less assistance is authorized to be provided under an emergency declaration in comparison to that authorized for a major disaster declaration. The types of emergency assistance authorized to be provided under an emergency declaration include the following:

- activities to support state and local emergency assistance;
- coordination of disaster relief provided by federal and non-federal organizations;
- technical and advisory assistance to state and local governments;
- emergency assistance through federal agencies;
- debris removal through grants to state and local governments (Section 407);
- grants to individuals and households for temporary housing and uninsured personal needs (Section 408); and
- distribution of medicine, food, and consumables.[86]

FUNDING CAPS AND COST SHARES

Before federal assistance is provided under Stafford Act authority contractual agreements between FEMA and the state must be finalized. These agreements set forth the terms under which FEMA will obligate and distribute disaster assistance funding to the state for redistribution to eligible applicants (e.g., local governments).

Pursuant to Stafford Act requirements, federal assistance programs are limited either to a fixed dollar amount or to a percentage of eligible costs. All parties use the agreements to establish federal, state, and local cost shares pursuant to statutory requirements and allowances. The Stafford Act stipulates that the minimum federal assistance for certain eligible activities "shall be not less than 75% of the eligible cost of such assistance."[87] Other provisions specify ceilings, rather than floors, for federal aid. Specific cost share requirements set out in the statute include the following.

- *Essential assistance:* The federal share must be at least 75% of eligible costs.[88]

- *Repair, restoration, or replacement of public facilities*: In general, at least 75% of eligible costs must be provided, but this threshold may be reduced to 25% if a facility has previously been damaged by the same type of disaster and mitigation measures have not been adopted to address the hazard. Federal aid generally will be reduced if facilities in flood hazard areas are not covered by flood insurance. Cost estimation requirements must be adhered to, but the President may approve costs that exceed the regulatory limitations. "Associated costs," such as the employment of national guard forces, use of prison labor, and base and overtime wages for employees and "extra hires," may be reimbursed. The President must notify congressional committees with jurisdiction before providing more than $20 million to repair, restore, or replace facilities.[89]
- *Debris removal*: The federal share must be at least 75% of the eligible costs.[90]
- *Individual and household assistance:* Temporary housing units may be provided directly to victims of disasters, without charge, for up to 18 months, unless the President extends the assistance "due to extraordinary circumstances." Fair market rents may be charged at the conclusion of the 18 month period.[91] The federal share of housing assistance is 100%. Financial assistance is also provided for Other Needs Assistance (ONA). This includes uninsured medical, dental, funeral, transportation, personal property, and other needs; the federal share for ONA is capped at 75% of eligible costs; the total amount that may be provided under the Individuals and Household Program (IHP) cannot exceed $25,000 (adjusted annually).[92]
- *Small project grants:* If the estimated costs of assistance for facility repair or replacement (Section 406), essential assistance (Section 403), debris removal (Section 407), or emergency assistance (Section 502) do not exceed $35,000 (adjusted annually), a small project grant may be issued.[93]
- *Emergency declaration assistance:* Federal assistance must constitute at least 75% of eligible costs. Expenditures made under an emergency declaration are limited to $5 million per declaration unless the President determines that there is a continuing need; Congress must be notified if the $5 million ceiling is breached.[94]

In instances where the state-local match is unduly burdensome due to factors such as the impact of the disaster or the fiscal condition of the state and its units of local government, the President may waive some or all of the cost-sharing requirement for PA programs. However, the President may not waive the 25% state-local requirement for assistance provided under the Individual

Assistance (IA) Program, except for insular areas.[95] As noted previously in this report, Congress statutorily waived the cost share provision for assistance provided after specified disasters.[96]

If a state or a local government believes that the economic impact from the disaster warrants, officials may contact FEMA to request a reduction in their portion of the federal cost-share. The regulations specify that an adjustment in the cost-share requirement may be made "whenever a disaster is so extraordinary that actual federal obligations under the Stafford Act, excluding FEMA administration cost, meet or exceed" specified thresholds.[97] (The costs incurred are based on Stafford Act obligations for the disaster and may differ from those of the preliminary damage assessment.[98])

HAZARD MITIGATION ASSISTANCE

Under the Hazard Mitigation Program Grant (Section 404 or HMGP) and the Pre-Disaster Mitigation grant (Section 203 or PDM) authorities the federal government provides grants to state and local governments to reduce the impact of future disasters.[99] HMGP funds are allocated to each state in which a major disaster has been declared. The funds generally may be used for any eligible hazard mitigation activity in the state (consistent with the Governor's request and Presidential designation), not necessarily related to the catastrophe that led to the declaration. PDM grants, by comparison, are applied for pursuant to criteria established in Section 203 and are not related to major disaster declarations.

Section 404 Hazard Mitigation Grants

HMGP funding was first authorized in the 1988 amendments to the 1974 statute.[100] Congress has amended Section 404 four times since 1988 to (1) increase the share of federal assistance and authorize property acquisition grants,[101] (2) stimulate state mitigation planning,[102] (3) reduce the maximum amount of federal aid,[103] and (4) establish a sliding scale for the awards.[104]

Money for HMGP derives from the Disaster Relief Fund (DRF, discussed in the next section of this report), not from line item appropriations. The current statutory provisions do not address what percentage would be used to determine hazard mitigation funding for disasters that exceed $35.333 billion. It would appear that special legislation would need to be enacted to provide mitigation funding for any disaster in excess of that amount.

Pre-Disaster Mitigation Grants

During the Clinton Administration FEMA initiated a new program titled Project Impact to stimulate pre-disaster mitigation efforts.[105] Congress appropriated funds, generally $25 million per year, in the mid-1990s, to support Project Impact activities. The 105th Congress considered, but did not approve, legislation to authorize funding for pre-disaster mitigation grants.[106] The issue carried over to the 106th Congress, which approved legislation to expand federal pre-disaster mitigation assistance in order to reduce federal disaster relief costs, save lives, and protect property through enactment of the DMA of 2000.[107]

The pre-disaster mitigation (PDM) grant program established by the DMA is intended to reduce losses and suffering "resulting from natural disasters" and provide a source of funding to ensure "the continued functionality of critical services and facilities after a natural disaster."[108] Assistance is authorized to help state and local governments implement "pre-disaster hazard mitigation measures that are cost-effective and are designed to reduce injuries, loss of life, and damage and destruction of property, including damage to critical services and facilities under the jurisdiction of the states or local governments."[109] Successful applicants must demonstrate the ability to develop "effective public-private natural disaster hazard mitigation partnerships. The legislation, according to FEMA, "emphasizes the importance of strong state and local planning processes and comprehensive program management at the state level."[110] Funding appropriated for PDM grants remains available until expended.[111]

DISASTER RELIEF FUND

Congress appropriates money to the Disaster Relief Fund (DRF) to ensure that the federal assistance described earlier in this report is available to help individuals and communities stricken by severe disasters.[112] Funds appropriated to the DRF remain available until expended. Such accounts are referred to as "no-year" accounts.[113] Appropriations to the DRF generally evoke little controversy. Supplemental appropriations measures are generally required each fiscal year to meet the urgent needs of particularly catastrophic disasters. While the hurricane season of 2009 (June through October) resulted in no major disaster declarations for named hurricanes, 56 severe storms during calendar year 2009 caused a level of destruction sufficient for the President to invoke Stafford authority.[114] In addition to the appropriation of $1.6 billion in regular appropriations for FY2010 to meet the needs associated with those and previous declarations, the fund also received an additional $5.5 billion through P.L. 111-112, based on a request from the Obama Administration for supplemental funding for FY2010 to pay for Gulf Coast recovery projects that have been the subject of decisions by an arbitration panel established pursuant to a congressional mandate.. The "Appeal Process" section at the end of this report discusses the impact of the panel's actions on the appropriation of supplemental FY2010 funds for the DRF.

Questions have been raised in the past concerning the increased cost of federal disaster assistance authorized by the Stafford Act as the categories of aid and eligibility for federal disaster assistance have expanded. For example, over the past five decades, Congress has expanded the basic authority first enacted in 1950 to include housing, grants for the repair of infrastructure, aid to individuals, loans to communities for lost revenue, and other needs. Congress has previously explored the issue of rising federal disaster assistance costs and reliance upon supplemental appropriations.[115] See "Reducing Federal Expenditures," for a discussion of this topic.

Mission Assignments

The DRF is the source of funding for "mission assignments" made by FEMA to other federal agencies. By this mechanism the federal government is able to provide assistance after a disaster by tasking any federal agency to undertake an activity necessary to save lives, protect property, or provide other assistance authorized by the Stafford Act. Mission assignments eliminate the need for Congress to appropriate specific amounts of money to many federal agencies. Instead, this instrument enables FEMA to task and reimburse other federal agencies for the assistance they undertake after the President issues a declaration.

For example, FEMA's Operation Blue Roof, the program that funds the installation of blue tarps to temporarily repair damaged residential roofs, is mission assigned to the U.S. Army Corps of Engineers (USACE). The Corps of Engineers obtains and distributes the tarps, hires the contractors, and ensures that all applicable federal regulations are followed. Other activities that may fall under FEMA mission assignments to other federal agencies include search and rescue, disease prevention and control, and health and medical support.

ISSUES FOR THE 112TH CONGRESS

As noted above, previously introduced legislation would amend provisions of the Stafford Act concerning eligibility for assistance, the types of assistance to be provided by the federal government, and administration of the emergency management process. Members of the 112th Congress might elect to consider the following in assessing whether further legislative changes are required.

Expanding Eligibility

The 1988 amendments to the Disaster Relief Act of 1974 added certain private nonprofit organizations that provide essential government services to the category of eligible recipients of Stafford Act public assistance.[116] Prior to that change, only state and local governments were eligible for the aid. Legislation introduced during the 111th Congress suggested further changes to the Stafford Act by enabling certain organizations or persons to receive assistance for which they are not currently eligible.

The most significant proposals would breach the long-standing bright line that has divided public and for-profit entities. Private for-profit organizations and property owners have not been deemed eligible for grants authorized by the statute since enactment of the 1950 statute and in subsequent amendments. Instead, Congress has directed that owners and operators of private, for-profit institutions must rely on pre-paid insurance or loans. The devastation caused by Hurricane Katrina to all components of many communities, and the recognition that community recovery is tied not only to public institutions but to private sector entities that provide essential services and employment, has revived interest in the issue of expanding eligibility for Stafford Act assistance to certain for-profit entities.

In addition to adding new organizations to the category of entities eligible for public assistance, some have suggested that the Congress take a different approach to extending Stafford Act assistance. Currently, only property owners are eligible to receive financial assistance for the repair or replacement of housing, some of the legislation from the 111th Congress sought to make several changes:

- H.R. 1059 would extend eligibility to the heirs of deceased property owners. Another approach is to modify the definition of the term "major disaster" to include other types of catastrophes.
- H.R. 1131 would authorize the President to issue such a declaration after instances of terrorist attacks, contamination, or most broadly, "or other catastrophic event."[117]
- Other bills (H.R. 3377 and H.R. 3453) would take a different approach; they would require that Administration officials revise and update the criteria currently used to determine whether a major disaster declaration will be issued.
- Also, one bill (H.R. 4297) would require that the President consider the needs of rural areas marked by less dense land development in determining eligibility for assistance. Revisions to the criteria for declarations could result in expanded eligibility for aid, or, as discussed below, possibly lead to reduced federal expenditures.

Some might contend that any such expansion of eligibility would result in considerably higher federal disaster relief expenditures as private, for-profit entities, newly designated property owners, or entire states would increasingly turn to federal grants in lieu of relying upon insurance policies or agreeing to repay loans. Accordingly, opponents of the proposals might argue that just as the 2000 amendments expanded certain federal assistance (largely through the establishment of a new pre-disaster hazard mitigation program) and reduced costs in some areas, the pending legislation should balance increased aid proposals with cost-cutting measures.[118]

Others would contend that amendments that would expand the boundaries of eligibility are appropriate as state and local governments struggle with budget problems, property owners face increased costs of credit, and policyholders face more expensive insurance premiums and more stringent insurance requirements. Also, they may contend that just as some of the facilities newly designated as eligible for public assistance funds in 1988, notably educational and health care institutions, provide services comparable to those available through public or non-profit organizations, the institutions identified in the pending legislation also provide public services critical to a community's revival. Such a conclusion, it might be argued, is consistent with congressional findings in the statute.[119]

As discussed below, precedent exists for the provision of such assistance. On at least two occasions, Congress appropriated funds to help private corporations deemed to be particularly affected after disasters—when utility companies' electricity lines were damaged in 1998 ice storms, and when other companies lost infrastructure in the terrorist attacks of 2001 in New York City. A third instance, involving the Texas Medical Center, directly involves Stafford Act eligibility criteria. The first two instances involved appropriations made outside of Stafford Act authority, the third consisted of special designation under the Stafford definition.

Private Utilities after the 1998 Ice Storms

In the early winter months of 1998, an ice storm resulted in the destruction of electricity distribution infrastructure as heavily laden trees collapsed on miles of poles and wires.[120] Private utilities owned and maintained the infrastructure. To address concerns that the private utilities would have to pass on the costs of repairs to customers, Congress included funding in the omnibus appropriations act for FY1999 (P.L. 105-277) for the Community Development Block Grant (CDBG) program administered by the Department of Housing and Urban Development (HUD). The act directed that funds would be provided "for disaster relief, long-term recovery, and mitigation in communities affected by Presidentially-declared natural disasters designated during fiscal years 1998 and 1999."[121] According to news reports, disagreement arose between the Secretary of HUD and Members of Congress over the use of the appropriated funds because the statute did not specify how the funds would be used.[122]

Private Utilities after the 2001 Terrorist Attacks

The destruction of much of the infrastructure around and under the World Trade Center on September 11, 2001, resulted in the appropriation of billions of dollars in federal assistance to New York City. The communications networks owned by private for-profit corporations were not eligible for assistance under the Stafford Act. Congress appropriated $783 million for a range of rebuilding efforts in Lower Manhattan, for economic revitalization and reconstruction in order to facilitate redevelopment, "including the restoration of utility infrastructure."[123] The conference report accompanying the legislation provided the following statements regarding this appropriation:

> The conference agreement includes an emergency appropriation of $783,000,000 for assistance to properties and businesses, including restoration of damaged infrastructure, and for economic revitalization activities in the areas of New York City affected by the September 11, 2001 terrorist attacks, instead of $750,000,000 as proposed by the House and Senate.
>
> The conferees recognize the tremendous human losses suffered by those businesses located in the World Trade Center, particularly those firms which suffered the greatest loss of life in the attacks. Because of the conferees' strong desire to support the redevelopment of the areas of New York City affected by the attacks and to encourage those businesses most devastated by the attacks to remain in New York City, the conferees have provided a $33,000,000 increase over the request. The conferees expect that these additional funds will be made available to assist those firms located in New York City at the time of the terrorist attacks which suffered a disproportionate loss of its workforce and who intend to re-establish their operations in New York City.
>
> The conferees concur with the language included in the House report encouraging the Lower Manhattan Development Corporation to consider the needs of utility companies and other institutions affected by the World Trade Center attacks.[124]

Private Health System after 2001 Tropical Storm Allison

In early June, 2001, President Bush issued a major disaster declaration for the state of Texas due to flooding associated with Tropical Storm Allison.[125] The Texas Medical Center (TMC), a health care complex that includes 13 hospitals as well as medical and nursing schools, suffered considerable damage.[126] The TMC, however, was not eligible for Stafford Act assistance because of its for-profit status. Through a provision included in an omnibus appropriations statute, TMC was declared eligible for Stafford Act assistance, as follows:

> That notwithstanding any other provision of law, for disaster declaration FEMA-1379-DR and hereafter, the Texas Medical Center is to be considered for FEMA Public Assistance and Hazard Mitigation grants as if it were an eligible applicant.[127]

Policy Options

Members of Congress might elect to take action on or in opposition to the legislation noted above in order to expand eligibility for Stafford Act assistance. Another option available to Members of Congress is to consider the integration of existing loan authority with the grant-in-aid provisions in the Stafford Act. Congress already moved in this direction in 2000 with the requirement that public assistance grants be available to certain non-profit facilities "only if" they provide critical services (defined in the statute) or the owner or operator has applied for a Small Business Administration (SBA) loan and has been found ineligible for such aid or has received the maximum amount of assistance possible through such a loan.[128] Congress might elect to consider modifying the thresholds for assistance by authorizing grants for businesses before they obtain loans at the maximum level or eliminating the requirement that they be found ineligible for such aid. One research organization cited the need for a legislative change such as this as follows:

> Currently, there are no federal grants programs — only loans — to assist small business. We further recommend converting some federal assistance for small businesses from loans to grants in order to ease the burden of loan repayments following a catastrophe. The amount of grants versus loans could be weighted depending upon the size and history of the business.[129]

The integration of the SBA disaster loan authority and the Stafford Act grant-in-aid assistance authority has previously been considered by Congress decades ago. First, when President Richard Nixon proposed the establishment of a Department of Community Development, the integration of "the disaster relief programs of the SBA and the Farmers Home Administration" were expected to be integrated with the transfer of emergency preparedness and disaster relief functions.[130] Congress did not act on President Nixon's proposal. Next, when President Jimmy Carter successfully reorganized federal emergency management authorities in 1978 through the creation of FEMA, the loan-making authority and resources of the SBA were viewed as separate activities from those moving into FEMA. In questioning Administration officials on the Carter proposal, one Senator argued for the separation of FEMA's coordination authority from the SBA loan authority as follows:

> I would think that, first of all, we would have for the first time in the government a single accountable public official with broad coordinative authority over all disaster relief programs. We are not going to be

> duplicating staffs in agencies such as the Farmers' Home Administration or Small Business Administration, who are very expert in making loans and following up and managing loans. But we are going to have a place in the government where overall policies can be developed ... I think we will be able to address overall emergency related policies, including those of programs that are actually carried out by other agencies.[131]

Members of the 112th Congress might elect to reopen the debate on whether loan authority, including that currently administered by the SBA, might be recast to address concerns raised in the discussion on whether Stafford act grant-in-aid eligibility might be expanded.

Reducing Federal Expenditures

The need for federal assistance after a disaster, particularly one of catastrophic magnitude, may lead some Members and other officials to promise "whatever it takes" to restore the area to its pre-disaster condition. The knowledge that families have been disrupted, livelihoods compromised, and basic governance threatened stimulate congressional action to ensure that past precedents are followed and, at times, additional aid is provided. Such assistance, however, requires the expenditure of revenues that might arguably be used elsewhere for other urgent purposes. For years Congress has wrestled with the competing goals of providing disaster assistance and controlling expenditures. The debate, often muted in the face of devastating losses, continues today.

In light of the ongoing discussions in the 112th Congress to reduce the federal deficit and federal expenditures, Members of Congress may wish to consider changes to the Stafford Act that would reduce federal expenditures. One of the most straightforward, and arguably difficult options, is to eliminate or reduce some of the benefits authorized by the statute. That option would require an examination of the components of the Stafford Act, their relative costs, and the effect of enacting changes to current benefits.[132] Should Members elect to consider other options, they may wish to consider past efforts to reduce expenditures.

Perhaps the most extensive effort to evaluate the range of federal disaster assistance programs and identify alternative financing approaches involved a Senate bipartisan task force. Following the high costs associated with Hurricane Andrew in 1992, Senators formed the Bipartisan Task Force on Funding Disaster Relief to compile information on federal policies and programs, identify costs, and discuss budgetary options. The final report issued by the Task Force included a discussion of one proposal to modify disaster assistance policies: establish more explicit and/or stringent criteria for providing federal disaster assistance.[133]

Criteria for Assistance Declarations

The Senate Bipartisan Task Force report focused on the criteria considered by the President in deciding whether a major disaster or emergency declaration would be issued. The policy that currently guides the decision process is set out in statute and regulations, and provides for discretionary judgment as well.[134] As some analysts have concluded, however, the criteria have not always been well documented or consistent.

Legislation introduced in the 111th Congress (H.R. 3377 and H.R. 3453) would have mandated updating of the existing regulatory criteria used to determine whether major disaster declaration assistance (and what type of assistance) is to be provided. Such an examination may lead to the tightening of requirements that could result in reduced federal expenditures. Conversely, less stringent requirements would result in expanded eligibility and assistance levels.

The issue of which criteria are used by the President to issue a declaration has been the focus of some analytical studies. Questions have been raised about whether federal aid is needed at all after some catastrophes. Some have contended that political considerations, as much as estimates of need, contribute to the costs of disaster relief as well. One 2002 study by economists Thomas A. Garrett and Russell S. Sobel purported: "States politically important to the president have a higher rate of disaster declaration by the president, and disaster expenditures are higher in states having congressional representation on FEMA oversight committees. Election year impacts are also found." Another study, which built upon the 2002 paper, examined presidential disaster declarations from 1981 through 2004 and found that "... the greater the electoral prize and the more competitive the statewide presidential contest, the more likely it is that a state will receive a presidential disaster declaration even after controlling for actual need."[135] Another, earlier perspective on the issue appeared in a 1989 study completed by Government Accountability Office (GAO) that also considered the effects of politics on disaster declarations. After examining presidential declaration data from the perspective of the party affiliation of governors and members of state congressional delegations, the authors concluded that there "were no indications that party affiliation affected White House major disaster declaration decisions."[136]

Members of Congress may elect to consider amending the Stafford Act to reduce the discretion exercised by the President in issuing the declarations. Such amendments might include establishing quantitative measures of need, damage thresholds higher than those currently used, or mandating that such declarations be subject to congressional review.

Individual and Family Eligibility

Stafford Act assistance for individuals (and the assistance authorized under the predecessor legislation since 1950) has never been means tested. All residents of areas included in major disaster or emergency declarations remain eligible for assistance, providing the President (through administrative officers) determine that the assistance is to be provided. Disaster survivors must provide an indication of need and assurance that federal aid will not be duplicative, but they do not have to meet income or other requirements.[137] It is arguable that the destruction of paper records and the need for victims to evacuate an area with scant or no proof of their income or other means of support, overwhelms any argument for means testing.

However, one piece of legislation introduced in the 111th Congress would have established increased individual and household assistance for disaster victims with annual incomes below $100,000. One might argue that if federal officials could obtain this information to assess the needs of lower income households for more assistance, similar information could be used to identify those with higher incomes in order to establish a maximum threshold for assistance.

In the face of record deficits and calls to reduce federal spending, Members of Congress may elect to evaluate the advantages and disadvantages of enacting a means test before housing or other needs assistance (referred to as ONA) are to be provided pursuant to Section 408 of the Stafford Act. Some may argue that means testing should not be used after a disaster when the full extent of losses is unknown, seemingly wealthy households will face new expenses, and administrative workloads will increase. Others may contend that such thresholds would result in lower federal expenditures, be consistent with the existing provision that federal aid not duplicative assistance from other sources, and encourage individuals and families to allocate personal resources to unanticipated needs.

Appeal Process

The most significant action taken by the 111th Congress regarding Stafford Act policy involved the authority of the U.S. Civilian Board of Contract Appeals to finalize decisions on public assistance applications. The FY2010 appropriations legislation for the Department of Homeland Security authorized the temporary reemployment of retired administrative law judges to preside over arbitration proceedings involving disputes of public assistance applications that exceed $500,000 in total costs.[138] Through this legislation Congress sought to expedite action on public assistance applications in the Gulf Coast that have been pending for years in order to stimulate recovery efforts in Louisiana, Alabama, Mississippi, and Texas associated with Hurricanes Katrina and Rita in 2005. Implementing regulations issued by FEMA outline the standard appeal process and summarize the conditions under which cases in appeal would be subject to the process established by Congress.[139]

From December 2009 through March 3, 2010, the Civilian Board of Contract Appeals issued 14 rulings on disputed applications.[140] The first five decisions set out the scope of the panel's authority.[141] Of the remaining nine decisions, one resulted in FEMA's favor (CBCA 1760),due to the lack of timeliness of the appeal, while seven were decided in favor of the appellants. The parties reached settlement in one decision (numbered CBCA 1781), obviating the need for a panel decision. Perhaps the most significant decision made during this period involved the disposition of Charity Hospital in New Orleans. In short, the panel ruled in favor of the appellants, and determined that FEMA should award almost $475 million for the replacement of the hospital (CBCA 1741).

Impact of Panel Decisions on Funding

The decisions to date by the appeals panel have influenced the debate between the Administration and Congress on appropriations for the DRF. Administration officials determined that the funds appropriated for FY2010 were insufficient for the Fund, in part due to the decision involving Charity Hospital and the expectation that other appeals will also result in increased demands for DRF assistance. On February 12, 2010, the President submitted a second supplemental FY2010 budget request for an additional $1.5 billion for the DRF, in addition to the $3.6 billion supplemental included in the FY2011 submission to Congress. This total supplemental of $5.1 billion, according to the Administration, would be expected to meet needs in the current fiscal year, including "arbitration panel decisions likely to impact the Disaster Relief Fund in a previously unexpected manner."[142] After Senate defeat on March 9, 2010 of a proposal to include the supplemental in legislation (H.R. 4123) that would extend expiring tax provisions, Members of Congress eventually passed the supplemental to the Disaster Relief Fund in P.L. 111-112 that provided more than $5.5 billion in additional supplemental disaster relief funding.

Summary of Issues on Appeals Process

While limited to appeals of decisions associated with Hurricanes Katrina and Rita, the congressional action indicates some degree of dissatisfaction with the standard appeals process established through regulation under the broad statutory guidance in the Stafford Act.[143] As opined by a former General Counsel for FEMA, the decision by Congress to authorize the arbitration process reflects "congressional belief that the appeal process for FEMA's public assistance program was broken ..."[144]

The issue before Congress is whether the appeal process requires a "fix," and whether that modification should take the form of a statutory change.

The regulations issued by FEMA implementing the arbitration process identified the criteria to be considered in determining whether public assistance applications could be submitted to the arbitration process as follows: large projects that exceed the $500,000 statutory threshold, applicants who remained eligible to file an appeal under the standing regulations (44 CFR 206.206) and withdrew their appeal, and those applicants that met specified time frames and deadlines. The regulation vested final determinative authority with the majority of arbitration panel members.[145] According to the statement accompanying the final rule, approximately 44 appeals were pending in Mississippi and Alabama (Hurricane Katrina), Texas (Hurricane Rita) and Louisiana (both hurricanes) when the agency published the regulation. Also, 2,188 projects were awaiting initial rulings and were estimated to be subject to the process. Accordingly, FEMA expected that 127 appeals would be forthcoming out of the 2,188 large projects eligible for the arbitration process [146] The actual number of appeals submitted to arbitration, however, are far below the estimate. According to the former General Counsel for FEMA, "More than a dozen applicants have elected arbitration ...."[147]

At least three topics may be explored by Members of Congress if they elect to debate the need for a statutory change.

- First, would an arbitration panel or other determinative party outside of FEMA be designated to participate in the appeals process, or should appeals continue to be heard and decided upon solely by FEMA administrators? It may be argued that only FEMA officials are most familiar with the details of public assistance grants, the application requirements, and comparable applications.[148] The cost estimates, circumstances unique to that particular disaster (such as the availability of skilled contractors and material), need for modifying initial damage projections, and other factors arguably require considerable knowledge of the public assistance grant program.149 Others may contend that this familiarity of the process and ownership of required information complicates efforts by appellants to even determine whether they should seek arbitration or appeal a FEMA decision, and may not meet the statutory mandate that rules "provide for the fair and impartial consideration of appeals under this section."150 For example, the quality of information available to appellants may be questioned, as summarized by the former FEMA General Counsel, "The rules assume that an applicant has the information and documentation required to establish eligibility at the outset of the arbitration process.... In fact, FEMA frequently is the only party with information that allows applicants to understand and respond to the disallowance of cost in an appeal (or an arbitration). And FEMA's responsiveness to FOIA [Freedom of Information Act] requests is notoriously slow — generally requiring a number of months.151 One option that Members might consider involves enacting an amendment that would mandate that FEMA maintain and provide applicable information to the appellants within the time frame established for an applicant deciding to file an appeal. Another option would maintain the current appeals process but provide for a final determinative authority outside of FEMA, perhaps through the permanent establishment of an arbitration panel comparable to that established by the 111th Congress, through the Federal Arbitration Association, or an arbitrator assigned to the matter through the Federal Arbitration Act (9 U.S.C. 1 et seq.)

- Second, the arbitration panel established pursuant to the American Recovery and Reinvestment Act of 2009 is solely authorized to consider appeals related to the Stafford Act public assistance program. As noted previously in this report, assistance is also provided to individuals and households for housing and uninsured needs, and the statute provides for appeal regulations that address decisions related to any assistance provided under the authority of Title IV.

- There is no provision for public review or consideration of the arbitration panel process. Congress might elect to consider whether the proceedings should remain closed.

End Notes

[1] The Robert T. Stafford Disaster Relief and Emergency Assistance Act, 42 U.S.C. 5121 et seq. "States" include U.S. territories and the District of Columbia. In addition to the assistance authorized by the Stafford Act, a wide range of disaster aid is provided by other federal agencies under statutory authority that specifically refers to disaster assistance, as well as under general assistance provisions. For information on the range of federal programs see CRS Report RL31734, *Federal Disaster Recovery Programs: Brief Summaries*, by Carolyn V. Torsell.

[2] See "Types of Assistance and Eligibility" in this report for information on the categories of aid available under Stafford Act authority.

[3] U.S. President George W. Bush, "Amendment of Executive Orders, and Other Actions, in Connection with the Transfer of Certain Functions to the Secretary of Homeland Security," E.O. 13286, Sec. 52, February 28, 2003.

[4] The authority of the FEMA Administrator to assist the President were incorporated into the Homeland Security Act, as amended by P.L. 109-295, 120 Stat. 1398 (6 U.S.C. 314(8)).

[5] 42 U.S.C. 5122(2).

[6] President Bush issued declarations in January 2009 before President Obama took office. For information on declarations see "2009 Federal Disaster Declarations," at http://www.fema.gov/news/disasters.fema?year=2009. For information on the declaration process see CRS Report RL34146, *FEMA's Disaster Declaration Process: A Primer*, by Francis X. McCarthy.

[7] Ibid.

[8] 42 U.S.C. 5187.

[9] See http://www.fema.gov/news/disasters.fema?year=2009.

[10] For current appropriations information see CRS Report R40642, *Homeland Security Department: FY2010 Appropriations*, coordinated by Jennifer E. Lake. For historical information see CRS Report R40708, *Disaster Relief Funding and Emergency Supplemental Appropriations*, by Bruce R. Lindsay and Justin Murray.

[11] Congress appropriates funds for the community disaster loan program separate from the DRF.

[12] The Disaster Relief Act of 1974 was enacted in P.L. 93-288, 88 Stat 143 et seq. The 1988 amendments (P.L. 100-707, 102 Stat. 4689 et seq.), which retitled the 1974 statute to its current designation, constituted the most far reaching amendments until those enacted in 2006 in the Post-Katrina Emergency Management Reform Act (P.L. 109-295, 120 Stat. 1444-1457. In addition, the Disaster Mitigation Act of 2000 (P.L. 106-390, 114 Stat. 1552 et seq. established some cost cutting requirements and authorized a new hazard mitigation program. For a history of the evolution of the statute and federal emergency management as a policy area see Rubin, Claire, ed. *Emergency Management; The American Experience, 1900-2005*, Public Entity Risk Institute (http://www.riskinstitute.org), 2007.

[13] The untitled 1950 statute is P.L. 81-875, 64 Stat. 1109-1111.

[14] In addition to amending the Stafford Act, the Post-Katrina Emergency Management Reform Act amended the Homeland Security Act to expand the authority of the Administrator of FEMA and the mission of the agency. See P.L. 109-295, 120 Stat. 1395-1406.

[15] P.L. 109-295, §681, 120 Stat. 1444, which amended Sec. 402 and 502 of the Stafford Act.

[16] P.L. 109-295, §688(3), which amended Sec. 102, 120 Stat. 1448 of the Stafford Act. See housing assistance at 42 U.S.C. 5174(d)(1)(A).

[17] P.L. 109-295, §689(b), which amended Sec. 403, 120 Stat. 1449 of the Stafford Act.

[18] P.L. 109-295, § 689(a), 120 Stat. 1448-1449. For more information see CRS Report RS22254, *The Americans with Disabilities Act and Emergency Preparedness and Response*, by Nancy Lee Jones.

[19] P.L. 109-295, §689a, 120 Stat. 1449.

[20] P.L. 109-295, §689e, new Stafford Act Sec. 616, 120 Stat. 1452. Note that Sec. 616 refers to the FEMA "Director," not Administrator, the title created in the 2006 amendments.

[21] P.L. 109-308, Sec. 3, Stafford Act amended Sec. 611, 120 Stat. 1725.

[22] P.L. 109-295, §689f, new Stafford Act Sec. 425 and Sec. 426, 120 Stat. 1452-1453. Note that the SAFE Ports Act also added a new Sec. 425 to the Stafford Act ("Essential Service Providers"), codified at 42 U.S.C. 4189e.

[23] For information on federal disaster housing assistance see CRS Report R40810, *FEMA Disaster Housing: From Sheltering to Permanent Housing*, by Francis X. McCarthy.

[24] P.L. 109-295, §689(c), Stafford Act amended Sec. 408, 120 Stat. 1449.

[25] P.L. 109-295, §689(c), Stafford Act amended Sec. 408, 120 Stat. 1449.

[26] P.L. 109-295, §686, Stafford Act amended Sec. 408(c), 120 Stat. 1448.

[27] The Stafford Act provides for annual cost of living adjustments to the funding ceiling. Effective October 1, 2006, the maximum amount that may be provided to an individual or household under the IHP authority is $28,800. See 71 *Federal Register* 59514.

[28] P.L. 109-295, §685, Stafford Act amended Sec. 408(c), 120 Stat. 1447.

[29] P.L. 109-295, §689d, Stafford Act amended Sec. 408, 120 Stat. 1452.

[30] P.L. 109-295, §689k, 120 Stat. 1456.

[31] P.L. 109-295, § 689i, 120 Stat. 1454. Following the Midwest floods in May of 2008 the state of Iowa made use of the pilot program, as did Texas following Hurricane Ike in the fall of 2008. FEMA has indicated it believes it has sufficient authority within the Stafford Act to continue this pilot program when needed.

[32] P.L. 93-288, 88 Stat. 154, amended by P.L. 100-707, 102 Stat. 4699. In 1988 Congress reduced the in-lieu grant from 90% to 75% of the federal estimate and retained the 90% level solely for areas with unstable soil (P.L. 106-390, 114 Stat. 1563).

[33] P.L. 109-347, §609, Stafford Act amended Sec. 406, 120 Stat. 1942.

[34] P.L. 109-347, §610, Stafford Act amended Sec. 407, 120 Stat. 1942-1943.

[35] P.L. 109-347, §608, 120 Stat. 1942.

[36] P.L. 109-295, §689j, 120 Stat. 1455. The statute terminated the program in December 2008. FEMA has expressed its intention to incorporate some of the PA pilot program through the rule-making process.

[37] P.L. 109-295, § 688(1), Stafford Act amended Sec. 102(9), 120 Stat. 1448.

[38] Ibid. The regulations are found at 44 CFR §206.221(e)(7).

[39] P.L. 109-295, § 689h, Stafford Act amended Sec. 406, 120 Stat. 1453.

[40] 42 U.S.C. 5170c.

[41] P.L. 109-295, §684, Stafford Act amended Sec. 404, 120 Stat. 1447.

[42] P.L. 109-295, §687, Stafford Act amended Sec. 302, 120 Stat. 1448.

[43] P.L. 109-295, §689g, new Stafford Act Sec. 326, 120 Stat. 1453.

[44] P.L. 109-295, §633, Stafford Act amended Sec. 303, 120 Stat. 1421. Note that the statute refers to the "Director" of FEMA, whereas the position is identified elsewhere as "Administrator."

[45] P.L. 109-347, §607, new Stafford Act Sec. 425, 120 Stat. 1941. Note that the Post Katrina Act also added a new Sec. 425 ("Transportation Assistance") to the Stafford Act.

[46] 42 U.S.C. 4102a 47 The Chair of the Transportation and Infrastructure Committee, Representative Thompson, sponsored H.R. 3437. Mr. Thompson has reportedly asked FEMA to halt the sale of trailers to the public, citing health hazards identified in congressional investigations. See Rob Margetta, "Experts Ponder Expanded Role for FEMA in Foreign Disasters," *CQ Homeland Security*, Jan. 14, 2010, at http://homeland.cq.com/hs/display.do?docid=3279089&sourcetype=31.

[48] This issue received attention in a hearing held during the first session of the 111th Congress. See U.S. House, Transportation and Infrastructure Committee, Subcommittee on Economic Development, Public Buildings, and Emergency Management, hearing, *Post Katrina: What it Takes to Cut the Bureaucracy*, July 27, 2009, transcript and statements at http://transportation.house.gov/hearings/hearingDetail.aspx?NewsID=960.

[49] Portions of this bill are comparable to H.R. 6658 ordered to be reported from the House Committee on Transportation and Infrastructure in the 110th Congress, but not considered by the full House nor the Senate.

[50] P.L. 111-32, 123 Stat. 1882-1883.

[51] P.L. 111-83, 123 Stat. 2185; P.L. 111-5, 123 Stat. 164.

[52] The text of the statute and pertinent regulations are presented in CRS Report RL33030, *The Budget Reconciliation Process: House and Senate Procedures*, by Robert Keith and Bill Heniff Jr. For further discussion of Stafford Act declarations, see CRS Report RL34146, *FEMA's Disaster Declaration Process: A Primer*, by Francis X. McCarthy.

[53] Whereas the first two pre-event actions are specifically authorized by the Stafford Act, the pre-event actions are inferred from general authority. Following an investigation into the response to Hurricane Andrew in 1992 the General Accounting Office (now the Government Accountability Office) reported that "Current federal law governing disaster response does not explicitly authorize federal agencies to undertake preparatory activities before a disaster declaration by the President, nor does it authorize FEMA to reimburse agencies for such preparation, even when disasters like hurricanes provide some warning that such activities will be needed." U.S. General Accounting Office, *Disaster Management: Improving the Nation's Response to Catastrophic Disasters* (Washington: July 23, 1993), p. 3.

[54] The statute reads "During the immediate aftermath of an incident which may ultimately qualify for assistance under this title or title V of this Act ... the Governor of the state in which such incident occurred may request the President to direct the Secretary of Defense to utilize the resources of the Department of Defense for the purpose of performing on public and private lands any emergency work which is made necessary by such incident and which is essential for the preservation of life and property. If the President determines that such work is essential for the preservation of life and property, the President shall grant such request to the extent the President determines practicable. Such emergency work may only be carried out for a period not to exceed 10 days." 42 U.S.C. 5170b(c).

[55] 44 CFR 206.35(c).

[56] Sec. 420 of the Stafford Act, 42 U.S.C. 5187.

[57] Regulations are found at 44 CFR 204.1 et seq.

[58] 44 CFR 204.51.

[59] 44 CFR 204.24.

[60] P.L. 109-295, §632, 120 Stat. 1421.

[61] For more information on the declaration process see CRS Report RL34146, *FEMA's Disaster Declaration Process: A Primer*, by Francis X. McCarthy.

[62] For criteria considered in the declaration of a major disaster, see 44 CFR 206.48. The gubernatorial request for a declaration is forwarded to the President through FEMA officials. Only the President may issue a major disaster declaration.

[63] 42 U.S.C. 5122(2).

[64] Each year FEMA issues a notice that identifies the threshold to be used as one factor to be considered in the determination of whether PA or IA or both will be made available after a major disaster declaration has been issued. The regulations establish a minimum threshold of $1 million in PA damages for each state; see 44 CFR 206.48(a)(1). Major disasters declared on or after October 1, 2005, would generally be expected to reach the threshold of $1.18 per capita for PA assistance to be authorized; see 70 FR 58734. However, the statewide threshold is not the sole factor. Assessments consider concentrations of damages in local jurisdictions even if statewide damages are not severe. Countywide impacts from major disasters declared on or after October 1, 2005, would generally be expected to reach the threshold of $2.94 per capita for PA assistance to be authorized; see 70 FR 58734.

[65] Citations to emergency assistance statutory authorities administered by agencies other than the Department of Homeland Security are identified in Table 2 of CRS Report RL33064, *Organization and Mission of the Emergency Preparedness and Response Directorate: Issues and Options for the 109th Congress*, by Keith Bea.

[66] Refer to the table in regulations (44 CFR 206.48(b)(6)) for computed averages of individual assistance needed for small, medium and large states, based upon losses incurred from 1994 to 1999.

[67] A Stafford Act "emergency" is "any occasion or instance for which, in the determination of the President, federal assistance is needed to supplement state and local efforts and capabilities to save lives and to protect property and public health and safety, or to lessen or avert the threat of a catastrophe in any part of the United States." 42 U.S.C. 5122(1).

[68] "The President may exercise any authority vested in him by ... this title with respect to an emergency when he determines that an emergency exists for which the primary responsibility for response rests with the United States because the emergency involves a subject area for which, under the Constitution or laws of the United States, the United States exercises exclusive or preeminent responsibility and authority. In determining whether or not such an emergency exists, the President shall consult the Governor of any affected state, if practicable. The President's determination may be made without regard to subsection (a) of this section." 42 U.S.C. 5191(b).

[69] 44 CFR 206.37(2).

[70] Sec. 402 of the Stafford Act, 42 U.S.C. 5170a.

[71] Sec. 403 of the Stafford Act, 42 U.S.C. 5170b. FEMA has established two categories of public assistance (PA) emergency work under Section 403 authority—debris removal (Category A) and emergency protective measures to save lives and property (Category B).

[72] Sec. 404 of the Stafford Act, 42 U.S.C. 5170c. The Hazard Mitigation Grant Program (HMGP), authorized by Section 404 of the Stafford Act (42 U.S.C. 5170c), funds activities that reduce the impacts of future disasters. For more information see CRS Report R40471, *FEMA's Hazard Mitigation Grant Program: Overview and Issues*, by Natalie Keegan. A second hazard mitigation program, the Pre-Disaster Mitigation (PDM) program, is authorized in Title II of the Stafford Act (42 U.S.C. 5133). For more information see CRS

Report RL34537, *FEMA's Pre-Disaster Mitigation Program: Overview and Issues*, by Francis X. McCarthy and Natalie Keegan.

[73] Sec. 405 of the Stafford Act, 42 U.S.C. 5171. The statute also mandates that federal agencies evaluate and take action to mitigate natural hazards.

[74] Sec. 406 of the Stafford Act, 42 U.S.C. 5172. Private nonprofit facilities that provide "critical services" (power, water, sewer, wastewater treatment, communications, and emergency medical care) may receive grants. Owners of other facilities that provide essential, but not critical services, must first apply for a Small Business Administration (SBA) loan, and may then receive grants if they are ineligible for such a loan or require aid above the amount approved by the SBA. The permanent work supported under this authority has been designated by FEMA as follows: "Category C," roads and bridges; "Category D," water control facilities; "Category E," buildings and equipment; "Category F," utilities; and "Category G," parks, recreational facilities, and other items. For more information see U.S. Department of Homeland Security, Federal Emergency Management Agency, "Public Assistance Guide-FEMA Publication 322," available at http://www.fema.gov/government/grant/pa/paguided.shtm, visited August 1, 2006.

[75] Sec. 407 of the Stafford Act, 42 U.S.S. 5173. Debris removal grants authorized by Section 407 are provided to states and are separate from the Category A debris removal PA assistance authorized by Section 403.

[76] Sec. 408 of the Stafford Act, 42 U.S.C 5174. [Sec. 409, food coupons and distribution, was redesignated Sec. 412.]

[77] Sec. 410 of the Stafford Act, 42 U.S.C. 5177. For background information see CRS Report RS22022, *Disaster Unemployment Assistance (DUA)*, by Julie M. Whittaker.

[78] 42 U.S.C. 5177a.

[79] Sec. 412 of the Stafford Act, 42 U.S.C. 5179.

[80] Sec. 413 of the Stafford Act, 42 U.S.C. 5180.

[81] Sec. 415 of the Stafford Act, 42 U.S.C. 5182.

[82] Sec. 416 of the Stafford Act, 42 U.S.C. 5183.

[83] Sec. 417 of the Stafford Act, 42 U.S.C. 5184.

[84] Sec. 418 of the Stafford Act, 42 U.S.C. 5185.

[85] Sec. 419 of the Stafford Act, 42 U.S.C. 5186.

[86] Section 502 of the Stafford Act, 42 U.S.C. 5192.

[87] The Stafford Act, as amended (P.L. 100-707), Sections 403(b), 403 (c)(4), 406(b), and 407(d). The same 75% minimum is also applicable to presidential emergency declarations (42 U.S.C. 5193, Section 503(a)). Neither the statute nor the regulations impose a formula for the distribution of cost-share requirements among state and local governments. Some states, for example, have established a 15% local and a 10% state match combination; other states equally divide the required cost share—12.5% each for the state and for each local government that receives Stafford Act assistance.

[88] 42 U.S.C. 5170b(b).

[89] 42 U.S.C. 5172. The statute provides that "base and overtime wages for the employees" and hires of state and local governments may be reimbursed under this authority (42 U.S.C. 5172(a)(2)(C). Regulations, however, specify that "straight- or regular-time salaries and benefits" of permanent employees (referred to as "force account labor costs") are not eligible if they are engaged in activities associated with essential assistance (Sec. 403) or debris removal grants (Sec. 407). See 44 CFR 206.228(a)(4).

[90] 42 U.S.C. 5173(d).

[91] 42 U.S.C. 5174(c). Financial assistance to build permanent or "semi-permanent" housing may be provided in insular areas outside the continental United States "and in other locations" where temporary housing alternatives are not available.

[92] 42 U.S.C. 5174. FEMA has established a limitation of $27,200 for IHP assistance to an individual or household in each emergency or major disaster declared after October 1, 2005. See 70 FR 58735. (The 2006 amendments deleted the caps on housing repair and replacement.)

[93] 42 U.S.C. 5189. FEMA has established a limitation of $57,500 for small project grants for major disaster or emergencies declared after October 1, 2005. See 70 FR 58735.

[94] Sec. 503 of the Stafford Act, 42 U.S.C. 5193.

[95] 48 U.S.C. 1469(a)(d).

[96] P.L. 111-32, 123 Stat. 1882-1883.

[97] 44 CFR 206.47(b).

[98] Due to the difficulty of conducting preliminary damage assessments immediately after a disaster occurs, it is not unusual for amendments to be made to the disaster declaration providing more assistance to affected areas once the greater degree of damage has become known and verified. For more information, see CRS Report R41101, *FEMA Disaster Cost-Shares: Evolution and Analysis*, by Francis X. McCarthy.

[99] For more information see CRS Report R40471, *FEMA's Hazard Mitigation Grant Program: Overview and Issues*, by Natalie Keegan and CRS Report RL34537, *FEMA's Pre-Disaster Mitigation Program: Overview and Issues*, by Francis X. McCarthy and Natalie Keegan.

[100] P.L. 100-707, Sec. 106(a), 102 Stat. 4698-4699. For more information see CRS Report.

[101] 1993 amendments in P.L. 103-181, §§2(a), 3, 107 Stat. 2054.

[102] 2000 amendments in P.L. 106-390, Title I, §104(a), 114 Stat. 1558.

[103] 2003 amendments in P.L. 108-7, Title IV, §417, 117 Stat. 525.

[104] 2006 amendment in P.L. 109-295, Title VI, Subtitle E, Sec. 684, 120 Stat. 1447.

[105] For background on Project Impact see U.S. Congress, House Committee on Transportation and Infrastructure, Subcommittee on Water Resources and Environment, *Disaster Mitigation, Preparedness and Response,* hearing, 105th Cong., 2nd sess., January 28, 1998 (Washington, GPO, 1988).

[106] See U.S. Congress, Senate Committee on Environment and Public Works, Subcommittee on Clean Air, Wetlands, Private Property and Nuclear Safety, *Federal Emergency Management Agency Reforms*, hearing, 105th Cong., 2nd sess., July 23, 1998 (Washington: GPO, 1999).

[107] Sec. 102, P.L. 106-290, 114 Stat. 1553-1557, 42 U.S.C. 5133.

[108] Sec. 101(b), P.L. 106-390, 42 U.S.C. 5133.

[109] 114 Stat. 1554.

[110] U.S. Federal Emergency Management Agency, "Hazard mitigation planning and Hazard Mitigation Grant Program," *Federal Register*, vol. 67, no. 38, February 26, 2002, pp.8843-54.

[111] See P.L. 108-334, the FY2005 Department of Homeland Security Appropriations Act, 118 Stat. 1313.

[112] Information on disaster funding is available in: CRS Report R40708, *Disaster Relief Funding and Emergency Supplemental Appropriations*, by Bruce R. Lindsay and Justin Murray.

[113] For background on this and other types of federal budget accounts, see CRS Report 98-410, *Basic Federal Budgeting Terminology*, by Bill Heniff Jr.

[114] The listing of declarations is found at http://www.fema.gov/news/disasters.fema?year=2009#sev1. Tropical storm/depression Ida, which did not reach hurricane status, led to declarations in two states.

[115] U.S. Congress, Senate Bipartisan Task Force on Funding Disaster Relief, *Federal Disaster Assistance*, S.Doc. 104-4, 104th Cong., 1st sess., (Washington: GPO, 1995). The House convened a task force that issued an unpublished report. Following completion of the task force efforts, some Members introduced a concurrent resolution (H.Con.Res. 39, 104th Congress) seeking a "fundamental overhaul of federal disaster policies." See also U.S. Congress, House Committee on the Budget, Task Force on Budget Process, *Budgetary Treatment of Emergencies*, hearing, 105th Cong., 2nd sess., June 23, 1998 (Washington: GPO, 1998).

[116] P.L. 100-707, 102 Stat. 4699. The 2000 amendments further modified the language concerning private nonprofit facilities by requiring that the facilities provide "critical services." See P.L. 106-390, 114 Stat. 1562.

[117] The phrase "other catastrophic event" is comparable to definitional language replaced in the 1988 amendments. From 1950 through 1988 the phrase "or other catastrophe" appeared in the definition of major disaster. (See 54 Stat. 1109 for the 1950 text.) In 1988 Congress modified the definition to more itemize the types of "natural" catastrophes that might result in a declaration and inserted "regardless of cause, any fire, flood or explosion" in place of the broader "other catastrophe" phrase (see 102 Stat. 4690 for the revised and current text).

[118] Title I of P.L. 106-390 authorized the pre-disaster mitigation program; Title II, "Streamlining and Cost Reduction," sought to reduce costs by authorizing the President to establish management cost rates, allowing states to administer their hazard mitigation (Section 404) grant programs, imposed restrictions on the awarding of grants to private nonprofit organizations, limited community development loans to $5 million, and made other changes.

[119] 42 U.S.C. 5121(b). The findings and declarations provision of the statute notes that the federal assistance is to, among other purposes, provide "programs for both public and private losses sustained in disasters." 42 U.S.C. 5121(b)(6).

[120] For background on the storm , see Senator Olympia Snowe, "The Ice Storm of 1998," remarks in the Senate, *Congressional Record*, vol. 144, February 11, 1998, S609-S612.

[121] 112 Stat. 2681-579. The full text of the provision that appropriated funds after the ice storms follows in the appendix to this memorandum.

[122] Steve Campbell, "Ice Storm Aid Pittance Seen as `Betrayal'," *Press Herald* online, November 22, 1998, at http://www.portland.com/sunews/story2.htm, visited November 23, 1998; Charles Davant, "Congress Hastens Maine Ice Storm Relief," *Bangor Daily News*, May 12, 1999; site archived.

[123] 116 Stat. 890.

[124] U.S. Congress, Conference Committees, 2002, *Making Supplemental Appropriations for Further Recovery from and Response to Terrorist Attacks on the United States for the Fiscal Year Ending September 30, 2002, and for Other Purposes*, conference report to accompany H.R. 4775, H.Rept. 107-593, 107th Cong., 2nd sess. (Washington: GPO, 2002), p. 180.

[125] Information is available at U.S. Federal Emergency Management Agency, "Texas Severe Storms and Flooding: Declared June 9, 2001," at http://www.fema.gov/news/event.fema?id=115, visited January 28, 2008.

[126] Summary information is available at Ellen Parson, "1,000-Year Flood Paralyzes Texas Medical Center," September 1, 2002, at http://www.ecmweb.com/mag/electric_year_flood_paralyzes/index.html, visited January 28, 2008.

[127] P.L. 108-7, 117 Stat. 514. No recorded debate exists relative to the merits of this provision.

[128] 42 U.S.C. 5172(a)(3)(A), enacted in P.L. 106-390, 114 Stat. 1562.

[129] Moss, Mitchell L. and Charles Shelhamer, New York University. *The Stafford Act: Priorities for Reform*. The Center for Catastrophe Preparedness and Response, 2007, p. 19, available at http://www.nyu.edu/ccpr/pubs/Report_StaffordActReform_MitchellMoss_10.03.07.pdf.

[130] U.S. Congress, House, Committee on Government Operations, Subcommittee on Legislation and Military Operations, *Reorganization Plan No. 1 of 1973*, hearing, Feb. 26, 1973, 93rd Cong., 1st sess., U.S. Govt. Print. Off., p. 37.

[131] U.S. Congress, Senate, Committee on Governmental Affairs, Subcommittee on Intergovernmental Relations, *Reorganization Plan No. 3 of 1978 (Disaster Preparedness)*, hearing, June 20, 1978, 95th Cong., 2nd sess., U.S. Govt. Print. Off., p. 8.

[132] Comprehensive data on federal emergency management funding are not collected on a regular basis. To the extent known, the most comprehensive collection of such data comprised expenditure information from fiscal years 1977 through 1993. A similar compilation has not been published. In constant dollars, federal disaster recovery costs totaled almost $87 billion, or slightly more than $5 million annually. The next most costly category involved hazard mitigation efforts, with a total of $27 million, or $1.6 million annually. Preparedness and response expenditures totaled less than $4 billion each for that period, or annual costs of less than $200 million. See U.S. Senate. Bipartisan Task Force on Funding Disaster Relief. *Federal Disaster Assistance, Report of the Senate Task Force on Funding Disaster Relief*, 104th Cong., 1st sess., Document No. 104-4 (Washington: GPO, 1995), Table 1.1, p. 5.

[133] U.S. Senate. Bipartisan Task Force on Funding Disaster Relief. *Federal Disaster Assistance, Report of the Senate Task Force on Funding Disaster Relief*, 104th Cong., 1st sess., Document No. 104-4 (Washington: GPO, 1995), p. 69. the report also discussed two other options: "emphasize hazard mitigation through incentives or rely more on insurance." The former is discussed earlier in this report as hazard mitigation authority exists in the Stafford Act. Greater reliance on pre-paid insurance policies raises issues beyond the scope of the Stafford Act.

[134] For information on the declaration criteria and procedures see CRS Report RL34146, *FEMA's Disaster Declaration Process: A Primer*, by Francis X. McCarthy.

[135] Andrew Reeves, "Political Disaster? Presidential Disaster Declarations and Electoral Politics," Abstract, unpublished paper, August 29, 2005, available at http://sws1.bu.edu/areeves/papers/fema.pdf.

[136] U.S. General Accounting Office, *Disaster Assistance: Timeliness and Other Issues Involving the Major Disaster Declaration Process*, GAO/RCED-89-138, May 25, 1989, pp. 1, 4.

[137] The establishment of means testing for individuals differs from the current policy used by the President and administration officials to establish criteria to determine whether declarations will be issued in states, and, if so, the considerations given to determine which counties will be included in the declaration. For information on such capacity and need measures see CRS Report RL34146, *FEMA's Disaster Declaration Process: A Primer*, by Francis X. McCarthy.

[138] The temporary reemployment authorization is found at Section 566, P.L. 111-83, 123 Stat. 218. The authorization for arbitration panels is found at Section 601, P.L. 111-5, 123 Stat. 164.

[139] U.S. Department of Homeland Security, Federal Emergency Management Agency, "Arbitration for Public Assistance Determinations Related to Hurricanes Katrina and Rita (Disasters DR-1603, DR-1604, DR-1605, DR-1606, and DR1607)," 74 *Federal Register* 44761-44769, August 31, 2009.

[140] The Board rulings are presented at http://www.cbca.gsa.gov/2009FEMA/2009FEMA.htm#October_2009.

[141] The first five rulings, found at the cite noted immediately above, are numbered CBCA 1739, 1741, 1775, 1783, and 1757.

[142] The White House, *FY2010 Supplemental Proposal in the FY2011 Budget*, p. 4 of the document found at http://www.whitehouse.gov/omb/assets/budget_ amendments/supplemental_02_12_10.pdf.

[143] Section 423 of the statute (42 U.S.C. 5189a) provides that any decision concerning Title IV (major disaster) assistance may be appealed within 60 days, requires that a decision be issued within 90 days, and that mandates that rules be issued "for the fair and impartial consideration of appeals under this section." The regulations allow eligible applicants, subgrantees, or grantees to first appeal a FEMA decision concerning an application for public assistance. Such appeals must be reviewed by grantees and then given to the FEMA regional director. If that appeal is denied, applicants may submit second round appeals to a FEMA Assistant Administrator for Disaster Assistance. The regulations set out time constraints on appellants and FEMA officials and authorize FEMA officials to, at their discretion, "submit the appeal to an independent scientific or technical person or group" with expertise in technical matters. "The decision of the FEMA official at the next higher level shall be the final administrative decision of FEMA." 44 CFR 206.206.

[144] Ernest Abbott, President, FEMA Law Associates, PLLC, *FLASH Newsletter*, vol. 4, Issue 11, April 2009, p. 1.

[145] U.S. Department of Homeland Security, Federal Emergency Management Agency, "Arbitration for Public Assistance Determinations Related to Hurricanes Katrina and Rita (Disasters DR-1603, DR-1604, DR-1605, DR-1606, and DR1607)," 74 *Federal Register* 44761-44769, August 31, 2009.

[146] The regulations note that FEMA funds slightly more than 5,000 large projects per year, and that almost 6% of those projects result in appeals. Ibid, p. 44765.

[147] Ernest Abbott, President, FEMA Law Associates, PLLC, *FLASH Newsletter*, vol. 4, Issue 13, December 2009, p. 1. Statistics on the number of projects submitted for arbitration have been requested from FEMA, but have not been provided as of the date of this report.

[148] In one decision the administrative judges questioned the quality of the investigative work undertaken by FEMA by including the following opinion of FEMA's substantiation of the denial for the replacement of Charity Hospital in New Orleans. "The FEMA representatives who testified at the hearing were less experienced and less credible than the BKA and RSMeans representatives. They had spent far less time in the building than had the BKA witnesses. Unlike the BKA and RSMeans witnesses, most of them were unlicensed. To the extent that FEMA relied on information in making the estimates on which version 3 of Project Worksheet 2175 was based, at least some of that information was incomplete; FEMA witnesses acknowledged that their estimates were subject to increase if FP&C presented appropriate proof that the facility's dimensions and disaster-related damages were greater than assumed by FEMA." CBCA 1741-FEMA, *In the Matter of State of Louisiana, Facility Planning and Control*, found at http://www.cbca.gsa.gov/2009FEMA/DANIELS_01-27-10_1741-FEMA__STATE_OF_LOUISIANA,_FACILITY_PLANNING_AND_CONTROL.pdf.

[149] Summary information on the requirements imposed on public assistance applicants is found at http://www.fema.gov/government/grant/pa/9580_5.shtm.

[150] 42 U.S.C. 5189a(c). The House committee report that accompanied the legislation containing this provision noted "The Committee intends that the President explore all available options to obtain a fair determination, including the consideration of alternative dispute resolution mechanisms to the extent they may be appropriate. The Committee intends to carefully oversee the implementation of this section." U.S. Congress, House, Committee on Public Works and Transportation, *Disaster Relief and Great Lakes Erosion Assistance*, 100th Cong., 2nd sess., H. Rpt. 100-517, March 15, 1988, p. 11. According to FEMA documentation available from the author of this report, the appeals provision derived from a proposal authored by then-Representative Tom Ridge; FEMA reportedly opposed the inclusion of the provision in the statute. Unpublished document by Arthur Bennett of FEMA, *Legislative History of P.L. 100-707 and Proposed Amendments to Disaster Relief Act of 1974, 1980-1987*, pp. 90-91.

[151] Ernest Abbott, President, FEMA Law Associates, PLLC, *FLASH Newsletter*, vol. 4, ssue 12, September 2009, p. 2.

In: Federal Emergency Management
Editor: Lewis B. Adams
ISBN: 978-1-62618-217-2

Chapter 4

FEMA'S DISASTER DECLARATION PROCESS: A PRIMER*

Francis X. McCarthy

SUMMARY

The Robert T. Stafford Disaster Relief and Emergency Assistance Act (referred to as the Stafford Act - 42 U.S.C. 5721 et seq.) authorizes the President to issue "major disaster" or "emergency" declarations before or after catastrophes occur. Emergency declarations trigger aid that protects property, public health, and safety and lessens or averts the threat of an incident becoming a catastrophic event. A major disaster declaration, issued after catastrophes occur, constitutes broader authority for federal agencies to provide supplemental assistance to help state and local governments, families and individuals, and certain nonprofit organizations recover from the incident.

The end result of a presidential disaster declaration is well known, if not entirely understood. Various forms of assistance are provided, including aid to families and individuals for uninsured needs and assistance to state and local governments and certain non-profits in rebuilding or replacing damaged infrastructure.

The amount of assistance provided through presidential disaster declarations has exceeded $140 billion. Often, in recent years, Congress has enacted supplemental appropriations legislation to cover

* This is an edited, reformatted and augmented version of a Congressional Research Service publication, CRS Report for Congress RL34146, from www.crs.gov, dated May 18, 2011.

unanticipated costs. While the amounts spent by the federal government on different programs may be reported, and the progress of the recovery can be observed, much less is known about the process that initiates all of this activity. Yet, it is a process that has resulted in an average of more than one disaster declaration a week over the last decade.

The disaster declaration procedure is foremost a process that preserves the discretion of the governor to request assistance and the President to decide to grant, or not to grant, supplemental help. The process employs some measurable criteria in two broad areas: Individual Assistance that aids families and individuals and Public Assistance that is mainly for repairs to infrastructure. The criteria, however, also considers many other factors, in each category of assistance, that help decision makers assess the impact of an event on communities and states.

Under current law, the decision to issue a declaration rests solely with the President. Congress has no formal role, but has taken actions to adjust the terms of the process. For example, P.L. 109-295 established an advocate to help small states with the declaration process. More recently, Congress introduced legislation, H.R. 3377, that would direct FEMA to update some of its criteria for considering Individual Assistance declarations.

Congress continues to examine the process and has received some recommendations for improvements. Given the importance of the decision, and the size of the overall spending involved, hearings have been held to review the declaration process so as to ensure fairness and equity in the process and its results.

BACKGROUND

Under the Robert T. Stafford Disaster Relief and Emergency Assistance Act (P.L. 93-288) there are two principal forms of presidential action to authorize federal supplemental assistance. Emergency declarations are made to protect property and public health and safety and to lessen or avert the threat of a major disaster or catastrophe.[1] Emergency declarations are often made when a threat is recognized (such as the emergency declarations for Hurricane Katrina which were made prior to landfall) and are intended to supplement and coordinate local and state efforts prior to the event such as evacuations and protection of public assets. In contrast, a major disaster declaration is made as a result of the disaster or catastrophic event and constitutes a broader authority that helps states and local communities, as well as families and individuals, recover from the damage caused by the event.[2] The differences between the two forms of declarations remain an area of study regarding what events may or may not qualify for the respective declarations.[3]

Federal disaster assistance has served as the impetus for many supplemental appropriations bills over the last several decades, and has accounted for presidential declarations in every state and territory.[4] The supplemental funds are placed in the Disaster Relief Fund (DRF) which is a "no-year" fund managed by the Federal Emergency Management Agency (FEMA) and used only for spending related to presidentially declared disasters.

Major disasters can be a dominant story in the mass media that captures attention both for the devastation that results as well as the potential help that is expected. As one observer noted:

> Disaster assistance is an almost perfect political currency. It serves humanitarian purposes that only the cynical academic could question. It is largely funded out of supplemental appropriations and thus does not officially add to the budget deficit. It promotes the local economy of the area where the building process occurs.[5]

While disaster assistance may be good "political currency," a disaster declaration is generally the result of a tragic and devastating incident that disrupts (and sometimes takes) the lives of hundreds or thousands of families and individuals and the communities and states where they reside. The long-term economic and environmental impact of a disaster can be severe. The assistance offered from federal and private sources may or may not be commensurate with the damage inflicted by a natural or man-made event. Following a disaster, years of rebuilding and recovery work may lie ahead for communities and states. It is the declaration process that sets the federal recovery help in motion.

The trigger for federal disaster assistance is contained in a relatively short statutory provision. P.L. 93-288 (the Stafford Act) includes one brief section that establishes the legal requirements for a major disaster declaration:

> Section 401. Procedures for Declaration. All requests for a declaration by the President that a major disaster exists shall be made by the Governor of the affected state. Such a request shall be based on a finding that the disaster is of such severity and magnitude that effective response is beyond the capabilities of the state and the affected local governments and that the federal assistance is necessary. As a part of such request, and as a prerequisite to major disaster assistance under this

> Act, the Governor shall take appropriate response action under state law and direct execution of the state's emergency plan. The Governor shall furnish information on the nature and amount of State and local resources which have been or will be committed to alleviating the results of the disaster and shall certify that, for the current disaster, state and local government obligations and expenditures (of which state commitments must be a significant proportion) will comply with all applicable cost-sharing requirements of this Act. Based on the request of a Governor under this section, the President may declare under this Act that a major disaster or emergency exists.[6]

The process for an emergency declaration is also contained in the Stafford Act. In part it is similar to a major disaster declaration in finding and process, but the actual authorities are limited. In addition, section (b) provides for a special authority for the President to exercise his discretion for events that have a distinctly federal character:

> Sec. 5191. Procedure for declaration. (a) Request and declaration All requests for a declaration by the President that an emergency exists shall be made by the Governor of the affected State. Such a request shall be based on a finding that the situation is of such severity and magnitude that effective response is beyond the capabilities of the State and the affected local governments and that Federal assistance is necessary. As a part of such request, and as a prerequisite to emergency assistance under this chapter, the Governor shall take appropriate action under State law and direct execution of the State's emergency plan. The Governor shall furnish information describing the State and local efforts and resources which have been or will be used to alleviate the emergency, and will define the type and extent of Federal aid required. Based upon such Governor's request, the President may declare that an emergency exists. (b) Certain emergencies involving Federal primary responsibility. The President may exercise any authority vested in him by section 5192 of this title or section 5193 of this title with respect to an emergency when he determines that an emergency exists for which the primary responsibility for response rests with the United States because the emergency involves a subject area for which, under the Constitution or laws of the United States, the United States and authority. In determining whether or not such an emergency exists, the President shall consult the Governor of any affected State, if practicable. The President's determination may be made without regard to subsection (a) of this section.[7]

The declaration process is elaborated upon in regulations, specifically in Subpart B of Part 206 (44 CFR). While these regulations have been adjusted through the regulatory process during the past three decades, since 1974 the procedures have undergone little significant change. The process itself is representative of the historical progression of federal disaster relief from being of an episodic nature to the current commonplace disaster declaration, now occurring on a weekly basis. The context in which disaster relief has grown has been in keeping with the growth of government and its concerns.

> Federal disaster relief has a long history in the U.S. dating back to the last years of the eighteenth century and arguably provided much of the political genesis for the New Deal social welfare programs (Landis 1999; Landis 1998; Moss 1999). As Michele Landis argues, social and political construction of claimants for relief as helpless victims of external forces beyond their control ("Acts of God") have exerted an enduring influence on American political discourse, which has manifested itself in heavy reliance on prior political precedents and analogies in constructing responses to current disasters.[8]

Information contained in the Preliminary Damage Assessments (PDAs) and summaries prepared by FEMA regional offices that accompany gubernatorial requests were long considered "pre-decisional and deliberative information" by the executive branch because they are part of the package that is developed and sent to the White House for the President's review and ultimate decision.

These materials have generally not been available under the Freedom of Information Act process.[9] However, during 2008, FEMA began to list on its website the results of the PDAs. While the summaries of FEMA recommendations are still not available, the agreed upon figures from the PDAs are now available to the public for review.[10]

Policy makers have found it difficult to achieve equity in the treatment of disparate natural disaster events. The events can vary widely in their type, scope, duration, and impact. Perhaps the greatest variables are the states they affect. Each state has a different topography, a different history, and different capacities to respond and recover based on their own authorities, resources, and choices in what they will do following a disaster.

Given those variations, it is a daunting task to construct a uniform process that can account for the range of natural and governmental circumstances that are a part of the nation's potential disaster landscape.

Congress and the Declaration Process

The Impetus for Reform

The question of how the federal disaster declaration process should work has come to the attention of Congress from time to time. Congress has requested reports from its investigative arms and investigated the process in panels that have considered aspects of the disaster relief process.[11] Various administrations have also considered the process and FEMA's role in it.

In 1981, and again in 2001, GAO issued reports on the declaration process that questioned the quality and consistency of FEMA's assessment criteria as well as the agency's ability to produce valid recommendations to the President on a governor's request for supplemental aid. As the most recent report (2001) concluded:

> These criteria are not necessarily indicative of a state's ability to pay for the damage because they do not consider the substantial differences in states' financial capacities to respond when disasters occur. As a result, federal funds may be provided for some disasters when they are not needed – a result that would be inconsistent with the Stafford Act's intent.[12]

Congressional interest in the declaration process derived, in part, from the increased cost in emergency spending—a recurring subject for the appropriations committees when considering supplemental spending legislation.[13] One review of disaster funding, particularly supplemental appropriations, found that the great majority of Disaster Relief Fund spending is related to the large, catastrophic events that meet the criteria established in the Stafford Act. These supplementals have been "driven" by the urgency of large natural disasters. As one witness explained in congressional testimony:

> Since funds provided in annual appropriations measures may be used to cover the federal costs of less catastrophic major disasters, Congress has focused critical attention on "must pass" legislation that would replenish the Disaster Relief Fund for the worst and most costly disasters.[14]

A review of data for a seven-year period from 1988 to 1995 reveals that large expenditures, as funded by supplemental bills, relate to declarations issued for the largest events.[15] During this time period, disaster declarations were made for Hurricane Hugo, the Loma Prieta earthquake, Hurricane Andrew, the Midwest floods of 1993, and the Northridge earthquake. However, these were not the only events deemed worthy of presidential action and of cost to the federal treasury. As summarized by one author:

> But like the tail of a comet, over 200 other declarations accounted for one quarter of such outlays, many of them of relatively minute cost and extent. While of lesser impact on the national treasury, such "low end" declarations have become, to some observers, new sources of federal spending at the local level, long referred to in other contexts as "pork barrel spending."[16]

During early 2001, the term "entitlement" was beginning to be used in describing federal disaster spending by the new Bush Administration. Underlining this thinking in his first appearance before the Senate Appropriations Committee, FEMA Director Joe M. Allbaugh explained:

> FEMA is looking at ways to develop a meaningful and objective criteria for disaster declarations that can be applied consistently. These criteria will not preclude the President's discretion but will help states better understand when they can reasonably turn to the federal government for assistance and when it would be more appropriate for the state to handle the disaster itself.[17]

During the 110th Congress, attention was given to the disaster declaration process and what states and local governments can reasonably expect from that process. During the early spring of 2007, there were tornadoes that had an impact on Arkansas, Alabama and Georgia. The latter two states received disaster declarations while the damage in Arkansas was deemed insufficient to warrant federal assistance. Following these actions, Chairman Bennie Thompson of the House Homeland Security Committee scheduled a hearing to review FEMA's procedures. Representative Thompson set the context as follows:

> As Members representing real communities back home, we want to understand just how FEMA makes determinations regarding what is a disaster deserving of attention and when folks have to fend for themselves. Quite simply, we must have a serious discussion on what our expectations are of our federal government and what should remain a state and local responsibility.[18]

In order to examine the process and the financial projections of disaster spending in particular, Congress mandated in P.L. 110-28 that GAO study how FEMA develops its estimates for supplemental funding that may be needed.[19] Report language detailed these instructions for the review:

> The Committee continues to be concerned with FEMA's ability to manage resources in a manner that maximizes its ability to effectively and efficiently deal with disasters. One aspect of particular concern is how FEMA makes projections of funding needed in response to any given disaster or to meet future disasters. A recent Government Accountability Office (GAO) report raised concerns about FEMA's ability to manage its day-to-day resources and the lack of information on how FEMA's resources are aligned with its operations. As a follow-up to this report, the Committee requests that within six months of enactment GAO review how FEMA develops its estimates of the funds needed to respond to any given disaster. Such review should include how FEMA makes initial estimates, how FEMA refines those estimates within the first few months of a disaster, and how closely FEMA's estimates predict actual costs. The review should also include additional analysis and recommendations regarding FEMA's ability to manage disaster-related resources in a manner that maximizes effective execution of its mission.[20]

The study parameters mandated by Congress are broader than the declaration process, placing its emphasis on ongoing FEMA spending, including the refinement and revision of early estimates as well as the general management of spending projections for the Disaster Relief Fund (DRF). But it recognizes the importance of the initial estimates in decision-making and that the declaration process represents the fundamental decision to both establish federal participation following a disaster and provides the initial estimated amount of resources needed that informs that decision.

GAO's eventual report in response to this request suggested that FEMA's estimates for disaster costs have improved but could use additional refinement; for example

> some of the factors that can lead to changes in FEMA's cost estimates are beyond its control, such as the discovery of hidden damage. Others are not, such as its management of mission assignments. Sensitivity analyses to identify the marginal effect of key cost drivers could provide FEMA a way to isolate and mitigate the effect of these factors on its early estimates. To better predict applicant costs for the Individual Assistance program, FEMA could substitute or add more geographically specific indicators for its national average.[21]

During the 111th Congress, a hearing before the House Homeland Security Committee focusing on FEMA's regional offices also elicited comments regarding the relative speed, and perceived lack of transparency, of the declaration process. The comments came from both a committee member and a state official closely involved in the process.

> Ranking Republican Mike D. Rogers of Alabama questioned the timeliness of FEMA's disaster declarations. Declarations that should take a few days often take a month or longer, and a lack of available information on the status of requests only adds to state officials' frustration, said Brock Long, director of Alabama's Emergency Management Agency. "When we make phone calls to the region or to headquarters, a lot of time the answer we get is 'the declaration request is in process' and that's it," Long said. [22]

The Skepticism of Reform

There have been several proposals to impose more stringent regulations on the declaration process in an attempt to more precisely assess disaster impacts, calculate eligible damage, and incorporate some measure of suffering and loss. Differing perceptions of the declaration process resulted in different reactions to these proposed reforms.

The history of the disaster declaration process is rife with reform efforts that were perceived by some not as reform but as punitive measures directed at certain constituencies. Some of those differing perceptions were held by those closest to the event at the local level that had experienced a disaster. A different perception was also often held by governors that wanted to protect their option to request federal help. These perceptions were also shared, in some instances, by Members of Congress representing affected areas. All elected officials argued, in various forms, that reform should not impede the delivery of needed federal aid.

One example, now in law, of the desire to reform but not obstruct the declaration process is in the Post-Katrina Emergency Management Reform Act of October of 2006. That statute created a new position at FEMA: the Small State and Rural Advocate.[23] One of the principal duties of the advocate is to ensure that the needs of smaller states and rural communities will be "met in the declaration process;" the advocate is also directed to "help small states prepare declaration requests, among other duties."[24]

Another example was FEMA's response to recommendations from GAO's 1981 report. FEMA drafted regulations in 1986 that would be more certain in their delineation of state requirements prior to a declaration, reduce overall federal contributions, and install a formula to determine whether a state would receive a presidential disaster declaration for repairs to state and local infrastructure (known as "public assistance" in the Stafford Act). As the congressional report on the initiative summarized the FEMA draft:

> The proposal would have limited the number of future presidential declarations by establishing a "state deductible" based on a per capita minimum dollar amount adjusted by the ratio of the state/local price index to the national index. Of 111 declarations issued in prior years, 61 would have been ineligible for any public assistance. FEMA also proposed to decrease the federal share for disaster costs from 75 percent to 50 percent and to exclude aid to special districts.[25]

The FEMA proposal of 20 years ago addressed some problems identified by students and critics of the declaration process. However, some Members of Congress viewed this proposal as a means of removing the power of discretion from elected leadership. Rather than apply the empirical solution suggested by FEMA to perceived problems in the declaration process, Congress instead legislated a provision to explicitly forbid the primacy of any "arithmetic formula." In place of agreeing to the regulatory changes in the formula proposed by FEMA and the Administration, Congress added the following section to the Stafford Act:

> Limitation on the Use of Sliding Scales. Section 320. No geographic area shall be precluded from receiving assistance under this Act solely by virtue of an arithmetic formula or sliding scale based on income or population.[26]

While enactment of this legislation halted FEMA's efforts, the issue did not disappear. As discussed below, FEMA has since adopted regulations that use, but not "solely", arithmetic formulae in determining need for assistance.

PRESIDENTIAL AND GUBERNATORIAL DISCRETION

The declaration process contains many factors for consideration and, for all but the most catastrophic events, the process moves at a deliberate speed accumulating information from several sources. While the process is informed by that information and its relationship to potential assistance programs, the information that is gathered at the state and local level does not preclude the exercise of judgment by the governor or the President.

The Stafford Act stipulates several procedural actions a governor must take prior to requesting federal disaster assistance (including the execution within the state of the state emergency plan and an agreement to accept cost-share provisions and related information-sharing). Still, the process leaves broad discretion with the governor if he or she determines that a situation is "beyond the capabilities of the state." The concession that a state can no longer respond on its own is difficult to quantify. It is the governor who makes that assessment, based on his or her knowledge of state resources and capabilities.

Actions by a governor are a driving constant in this process. Both declarations of major disaster and declarations of emergency must be triggered by a request to the President from the governor of the affected state.[27] The President cannot issue either an emergency or a major disaster declaration without a gubernatorial request. The only exception to this rule is the authority given to the President to declare an emergency when "he determines that an emergency exists for which the primary responsibility for response rests with the United States because the emergency involves a subject area for which, under the Constitution or laws of the United States, the United States can exercise exclusive or preeminent responsibility and authority."[28]

The importance of governors in the process is not lost on, nor would it likely be diminished by, Presidents who formerly served in that position. Over the last three decades of Stafford Act implementation, four of the Presidents during this period were former governors who had worked through the disaster declaration process from both the state and the federal level.[29]

Having that experience may have left the Presidents, and their staffs and appointees, with an appreciation of the discretionary authority inherent in the process. While there are some established standards in the law, the factors that are to be weighed in considering the impact of a disaster on the need for assistance for families and individuals are general considerations that underscore the judgment required to reach a decision. As one observer noted of the decision for general, flexible considerations:

> In other words, Congress likes to keep the process imprecise, even if benefits occasionally go to the undeserving. The absence of objective criteria preserves wide political discretion to the president.30

Congress also has among its number former governors who have exercised this discretion at the state level. Still, despite the interest some may have in keeping "the process imprecise," some Members of Congress express disappointment at times with the exercise of the discretion and the general nature of the considerations. As noted previously in this report, that disappointment has been reflected in hearings that have focused on how the disaster declaration process works in practice.

In the declaration process, FEMA develops a recommendation that is sent to the White House for action. However, as implied, it is a recommendation from FEMA and the Department of Homeland Security (DHS). The final action, as defined in the regulations of the process, is a "Presidential determination."[31] Just as the governor retains the discretion to request federal assistance regardless of thresholds or indicators, the President retains the discretion to make a decision that may be counter to recommendations he receives.

PRELIMINARY DAMAGE ASSESSMENTS

Although not explicitly mentioned in the Stafford Act, Preliminary Damage Assessments (PDAs) are a crucial part of the process of determining if an event may be declared a major disaster by the President. The minimal discussion of PDAs in the public record stands in inverse proportion to their impact on disaster decisions and subsequent expenditures from the Disaster Relief Fund (DRF).

When a PDA is conducted after an event it is the "mechanism used to determine the impact and magnitude of damage and the resulting unmet needs of individuals, businesses, the public sector, and the community as a whole."[32] The most "preliminary" part of a PDA may be an abbreviated one completed only by the state to determine if the situation merits development of a complete PDA with federal participation. Based on their previous experience, states may determine that the event will not reach the level where a federal disaster declaration is likely. However, despite findings that federal aid may not be needed, there may be political considerations that could lead to a gubernatorial request. As one author points out in his study of the process:

> Governors also feel the heat of media coverage of incidents in their states, and they too appreciate the importance of exhibiting political responsiveness. Governors also appreciate that their future political

> fortunes may be influenced by how they handle their disaster and emergency incidents. As a consequence, the hypothesis assumes that governors are the pivotal and decisive players in securing presidential disaster declarations and that they have a tendency to request declarations for even marginal events.[33]

While media and political pressure may have some influence on the outcome of some requests, governors may exercise caution since they are reluctant to be turned down when requesting aid. A denial of their request could be perceived by some to reflect adversely on their decision-making skills and judgment under pressure. Unlike the procedures of the federal process, a governor's decision to request a declaration is a public and often newsworthy action.

Regardless of any question regarding motivation, the governor's first decision is whether the incident is severe enough to assemble a traditional PDA team to survey the damaged area. The traditional PDA team includes a state official, representatives from the appropriate FEMA regional office, a local official familiar with the area and, in some instances, representatives from the American Red Cross and/or the Small Business Administration. The FEMA representatives have the responsibility of briefing the team on the factors to be considered, the information that will be helpful in the assessment and how the information should be reported. One significant improvement in this process is that the regulations now require that the participants reconcile any differences in their findings.[34]

Another factor is the quality of the PDA team and its findings. PDAs are ordered up quickly after an event. FEMA's 10 regional offices are often engaged in multiple disasters and have to rely on temporary employees (albeit, usually experienced ones) to staff the PDA team.[35] Also, given the variables among regions in interpreting policy and guidance, consistency of approach may also be a question. This became a significant issue during the hurricane season of 2004 in Florida, where questions were raised regarding the designations of some counties.[36]

When working openly and with federal and state cooperation, the PDA can be an effective and inclusive process. As Washington State Emergency Management Director Jim Mullen, and FEMA Region X Director Susan Reinertson, explained the process for a survey of flood damage in November of 2006:

> 'Joint PDA teams will visit and inspect damaged areas, document damage and talk, as needed, with homeowners and local officials,' said

> Mullen. 'It's a partnership effort designed to provide a clear picture of the extent and locations of damage in counties that have reported the most substantial damage to primary homes and businesses.'
>
> 'The PDA teams look at the total scope of damage to establish if recovery is beyond the capabilities and resources of the state and local governments. The PDA doesn't determine the total cost of recovery, nor does it guarantee a presidential declaration for individual assistance,' said FEMA Regional Director Susan Reinertson.[37]

PDA teams often face challenges in the collection of data. Some information may be observable in a survey of the area, such as the number of bridges damaged or the number of culverts washed out. But other necessary information, such as the percentage of elderly residents in an area, or the amount of insurance coverage for all homeowners or renters may be more difficult to obtain. Also the geographic span of the damage can create complications as the PDA team struggles to cover all of the affected area in a limited time. Further complications may then ensue based on how much of the area that the PDA teams have visited (generally counties) are included in the governor's request.

While PDAs are the usual way damages are assessed, there are exceptions to this rule. Some incidents are so massive in their scale and impact that the actual declaration is not in doubt. This would include events such as Hurricanes Andrew and Katrina, the Loma Prieta and Northridge earthquakes, the Mount St. Helens volcanic eruption, and the September 11 terrorist attacks. In these instances, the decision is not whether a declaration will be made, but how broad the coverage will be, both geographic and programmatic. In such cases the President, in accordance with the regulations, can waive the PDA requirement. But as the regulations note, a PDA may still be needed "to determine unmet needs for managerial response purposes."[38] This means the PDA helps to identify a specific, potential need for certain programs, such as crisis counseling or disaster unemployment assistance during the disaster recovery period. It is this identification of discrete need that helps the governor decide on which assistance programs will be requested.

The PDA is a "bottom up" process as information gradually rises up for decision-makers to consider. Multiple pressure points, including affected citizens, elected officials, and professionals in various fields, may all urge PDA team members to reach certain conclusions. As one author suggests:

> The president, motivated by the need to appear highly politically responsive, solicits and encourages a gubernatorial request for a

> presidential disaster declaration. Publicity is a factor in that CNN and other news organizations help to promote nationally what would otherwise be a local incident addressed by subnational authorities. The president may also be influenced by the electoral importance of the state that experiences the incident.[39]

It is difficult to overstate the importance of the media context as noted above. Depending on the news of the day, the disaster event in question may be the biggest national story and thus create momentum for action that is difficult to assuage with explanations of traditional administrative procedures.

Factors Considered for Public Assistance in Major Disaster Declarations

Public Assistance (PA) refers to various categories of assistance to state and local governments and non-profit organizations. Principally, PA covers the repairs or replacement of infrastructure (roads, bridges, public buildings, etc.) but also includes debris removal and emergency protective measures which cover additional costs for local public safety groups incurred by their actions in responding to the disaster.[40] In assessing the degree of PA damage, FEMA considers six general areas:

- Estimated cost of the assistance
- Localized impacts
- Insurance coverage
- Hazard mitigation
- Recent multiple disasters
- Programs of other federal assistance.

Estimated Cost of the Assistance

Although all of these factors are considered, the *estimated cost of the assistance* is a key component and may be "more equal" than other factors since it contains a threshold figure. What is especially noteworthy about the cost estimates is that they could be interpreted as an example of the "arithmetic formula" that was precluded from use in Section 320 of the Stafford Act. However, that section does state that a formula cannot be "solely" determinative of the fate of a governor's request. Use of the other factors listed arguably justifies FEMA's consideration of the threshold figure of $1 million in PA damage—the first number FEMA expects to see in a request that includes PA as an area of needed assistance. In addition FEMA also considers a state-wide threshold of $1.30 per capita in estimated eligible disaster costs before it will approve a request for PA help.[41]

Depending on the state's population, the per capita threshold may be difficult to reach. For example, the 2000 Census estimated California's population at just under 34 million people.[42] Applying the $1.30 per capita figure, it would require eligible PA damage in California to be close to $41.5 million. California is a large state with a budget and tax base commensurate with its size. Expecting a large state to be able to respond on its own is equating such help, and such amounts, to be within "the capabilities of the state."[43]

Compare that level of eligible damage for California ($41.5 million) with Nevada, a small population state according to FEMA regulations.[44] For Nevada, with a population of just under 2 million people according to the 2000 Census, eligible PA damage of about $2.4 million would make the state potentially eligible for supplemental federal assistance.

There are obvious differences in the populations of these two states; however, both have substantial industries and are growing areas. They are bordering states that are both subject to the threat of earthquake damage.[45] Different measurements have been suggested to try to more accurately capture a state's capacity to respond to disaster events. Some of these measurements are discussed in the "Congressional Considerations for the Declaration Process" section of this report.

Localized Impacts

The next factor used by FEMA to consider a declaration that may include PA help focuses on *localized impacts*. FEMA generally looks for a minimum of $3.27 per capita in infrastructure damage in a county before designating it for PA funding.[46] While such a per capita amount would likely ensure that the local entity (almost always a county) would be included in the list of jurisdictions designated for PA assistance should a declaration be issued, high local levels of damage per capita would not supersede a finding that damages state-wide fell below the statewide threshold amount. However, knowledge of a very large localized impact could weigh on the President's discretion and raise the importance of the *localized impact* factor.

Insurance Coverage

Insurance coverage also is considered in assessing damage. Officials preparing PA estimates deduct the amount of insurance that should have been held by units of governments and nonprofit organizations from the total eligible damage amount. However, this is a complicated assessment with several caveats. A considerable number of states and local governments "self-insure" their investments against some types of disasters and may argue that they would have total liability absent a federal disaster declaration.[47] Also, in the event of some disasters (earthquakes being a prime example), the state insurance commissioner must certify that hazard insurance is both available and affordable. If it is not available, deducting an amount of coverage not realistically available would not be a practical consideration.[48]

In the case of a flood event, it is much simpler for FEMA to access information available on whether flood insurance was available, given that FEMA administers the National Flood Insurance Program (NFIP) and there are program staff in each FEMA region who monitor NFIP participation. This knowledge of flood insurance availability and costs makes it a simpler process for FEMA to deduct the amount of flood insurance that should have been in place from the total potential award for a public structure.[49]

Hazard Mitigation

Hazard mitigation presents a challenging and different type of consideration for Public Assistance disaster declaration requests. If the requesting state can prove that their per capita amount of infrastructure damage falls short due to mitigation measures that lessened the disaster's impact, FEMA will consider that favorably in its recommendation to the President. Calculations of savings would be dependant on cost-benefit analysis and other related estimates of damages that were avoided. This factor is intended to encourage mitigation projects by states to lessen the risks of future natural disasters. Some might consider this practice to be one where "a good deed should go unpunished." However, another factor to be considered is that the mitigation work that is credited to the state may, in fact, have been principally financed (up to 75% of the costs) with previous FEMA disaster assistance funding through the Hazard Mitigation Grant Program (HMGP).[50]

Recent Multiple Disasters

Recent multiple disasters is the factor FEMA considers when a state has been repeatedly hit by disaster events (either presidential declarations or events within the state that were not declared) within the previous 12 months. FEMA evaluates the amount of funds that the state has committed to these recent events and their impact on the state and its residents. For example, a request from a state that has responded on it own to a series of tornadoes may receive a more favorable consideration, even if the catalyst for the request was arguably not as destructive as others. This factor was used recently in FEMA's assessment of the multiple hurricanes that struck Florida during the 2004 hurricane season.

Other Federal Programs

When FEMA is reviewing a governor's request it is also considering whether *other federal programs* are available. One example might be if a large amount of the reported damage occurred to federal-aid-system roads. The Federal Highway Administration (FHWA) is a more appropriate agency to handle such an event, and administers programs that address this specific type of damage.[51] The bridge collapse in Minnesota in August of 2007 was a dramatic example of this type of event that would warrant significant, non-FEMA, federal aid. Similarly, other federal programs may be more responsive to certain types of natural events and the problems they create. For example, oceanic bacteria could cause harmful failures for the fishing and hatcheries industry. Should such outbreaks occur, the Magnuson-Stevens Fisheries Act (P.L. 94-265) authorizes programs under the Commerce Department that would have more appropriate forms of emergency assistance for commercial fishermen than FEMA through the Stafford Act.[52] Also, while drought is a type of disaster that could result in a major disaster, such catastrophes usually result in program assistance from the U.S. Department of Agriculture rather than presidential declarations.[53]

In summary, all of these factors are used by FEMA to determine whether a major disaster declaration will be recommended and whether PA aid will be extended as a part of that potential declaration. FEMA's assistance to a homeowner who chooses not to purchase insurance is left with help from FEMA that is capped at $28,200. The assistance to repair public infrastructure is based on the amount of eligible damage caused by the disaster event. There is no cap, and the sums can be in the billions of dollars.

Factors Considered for Individual Assistance in Major Disaster Declarations

Individual Assistance (IA) includes various forms of help for families and individuals following a disaster event. The assistance authorized by the Stafford Act can include housing assistance, disaster unemployment assistance, crisis counseling and other programs intended to address the needs of people. In seeking to assess the impact of a disaster on families and individuals, the factors FEMA considers include:

- Concentration of damages
- Trauma
- Special populations
- Voluntary agency assistance
- Insurance coverage
- Average amount of Individual Assistance by state.

Concentration of Damages

Concentration of damages looks at the density of the damage in individual communities. FEMA's regulations state that highly concentrated damages "generally indicate a greater need for federal assistance than widespread and scattered damages throughout a state."[54] Concentrated damages are far more visible and may be an indication of significant damage to infrastructure supporting neighborhoods and communities, thereby increasing the needs of individuals and families.

However, the dispersion of damage is not necessarily an indication that individual and family needs are non-existent. Damage in rural states, almost by definition, is far less concentrated and could arguably be more difficult for a PDA team to view and assess. Congress has sought to address this challenge through the creation of the Rural and Small State advocate position at FEMA.[55]

Trauma

Trauma is defined in three ways in FEMA's regulations: the loss of life and injuries, the disruption of normal community functions, and emergency needs that could include an extended loss of power or water.[56]

Despite their prominence and importance to victims, families and communities, the loss of life and injuries have relatively little bearing on a declaration decision, but they would greatly influence media coverage that can influence the decision-making process. An extreme amount of losses would be traumatic and considered in the evaluation of the governor's request. But the actual help available from FEMA in response to such losses is limited. The Other Needs Assistance (ONA) program, a part of the Individuals and Households Program (IHP), may provide assistance for uninsured funeral expenses and medical help.[57] As noted, the extreme loss of life or many injuries will likely influence the amount of media coverage for the event. As previously explained in the "Presidential and Gubernatorial Discretion" section of this report, the media coverage can influence not merely the pace of the decision, but the actual decision itself.[58]

The other areas (disruption of functions and loss of utilities) in the trauma rubric can be a point of dispute in declaration decisions. What constitutes a disruption of normal community functions? Do road closures that result in the closing of schools equate to a disruption? Another consideration is how long the disruption remains and how, or even if, federal help can alleviate the disruption. Similarly, the emergency needs due to the loss of utilities are defined by the length of time the power or water are not in service. Because characteristics of these factors are not defined in regulations, discretionary judgments are significant aspects of the evaluation of IA needs.

Special Populations

Special populations are considered by FEMA in assessing a request for Individual Assistance.

FEMA attempts to ascertain information about the demographics of an area affected by a disaster event. Those demographics include the age and income of residents, the amount of home ownership in an area, the effect on Native American tribal groups, and other related considerations to be taken into account. The knowledge of the demographics within the affected area gives FEMA added information to consider regarding trauma and community disruption. Special populations are a factor in the consideration of a governor's request, but the total number of households affected that would be eligible for Stafford Act programs remains the prime consideration.

Voluntary Agency Assistance

Voluntary agency assistance involves an evaluation of what the volunteer and charitable groups and state and local governments have already done to assist disaster victims as well as the potential help they can offer in the recovery period. FEMA also considers whether state or local programs "can meet the needs of disaster victims."[59] This factor is among the most contentious in the disaster declaration process due to the subjectivity of the assessment.

FEMA's evaluation of local and state capabilities, and the capabilities of the local voluntary community, may vary greatly among catastrophes. There is an expectation that the regional office's relationship and history with the state could provide some of this information. But a cursory look at state programs available to "meet the needs of disaster victims" suggests that few resources are comparable to federal help in intent or in scope. A National Emergency Management Association (NEMA) survey of its members for state-funded disaster assistance showed that relatively few states provide assistance beyond that authorized in the Stafford Act.

NEMA noted that the "other" category could include a range of programs or funds not necessarily of direct aid to victims, including "local government loans, HAZMAT (hazardous materials) funds, disaster unemployment insurance, and a governor's contingency fund."[60.] The report does not define these programs any further, but the small numbers of programs directed at IA or unmet needs suggests that FEMA officials evaluate this area carefully given the lack of information regarding available resources for these programs and the extent of their coverage.[61]

In the same vein, attempts to assess the capacity of local voluntary and charitable groups to handle "unmet needs" caused by a disaster can be challenging and problematic. Part of the challenge is discerning the assistance available from the non-profit, voluntary sector, and if that aid meets the needs created by the disaster event.

Table 1. Summarizes data from that survey

Assistance Provided	Number of States with These Programs
Public Assistance Program	22
Individual Assistance Program	9
Unmet Need Program	1
Other	5

Source: NEMA 2010 Biennial Report: Organizations & Funding for State Emergency Management and Homeland Security (Lexington, KY: Council of State Governments) p. 4.

A problematic aspect of the assessment is similar to the "good deeds" concept addressed in the "hazard mitigation" factor in PA. In that instance, the criteria reward the mitigation work that communities and states had undertaken to lessen the impact of a disaster. Similarly, a strong and efficient charitable sector at the local level that is equipped and funded to address the remaining needs could result in a disaster not being declared and federal supplemental funding not being made available.

Arguably, a community does not base all of its preparedness decisions on the potential of FEMA funding in an extraordinary situation. However, it has been argued that some aspects of the FEMA declaration process could be viewed as disincentives for a sound, local capacity to deliver such assistance. One analyst believed that this problem underlines the need for clear criteria for smaller disaster events:

> Too low a threshold reinforces the perception that the federal government will always come like the cavalry to rescue states and local governments from their improvident failure to prepare for routine disasters. Lapses in preparedness, response, recovery, and mitigation (to cite the disaster management litany) should not be encouraged by a too readily available bailout by the federal government and taxpayers.62

In evaluating the capabilities of local organizations, FEMA seeks to determine the help that actually exists, rather than assistance that might be expected to be in place.

Insurance Coverage

As noted earlier in the PA section, *insurance coverage* is also an important consideration when FEMA considers a request for Individual Assistance. This is in part derived from the general prohibition in the statute of the duplication of benefits, as follows:

> ... each federal agency administering any program providing financial assistance to persons, business concerns, or other entities suffering losses as a result of a major disaster or emergency, shall assure that no such person, business concern or other entity will receive such assistance with respect to any part of such loss as to which he has received financial assistance under any other program or from insurance or any other source.[63]

This provision does not necessarily result in delayed assistance. FEMA is able to provide help to individuals and households that have disaster damages but are waiting on insurance or other assistance for help. Those applicants can receive FEMA help as long as they agree to reimburse FEMA when they receive their other assistance. If a disaster occurred where almost all of the damaged dwellings were fully insured for the damage that was sustained, FEMA could conclude that a disaster declaration by the President was not necessary. Among the types of disasters FEMA frequently responds to, tornado disasters particularly reflect this challenge since tornado coverage is a part of most homeowners insurance policies.

Since the NFIP is administered by FEMA, officials can quickly determine the status of flood insurance in communities and the number of policies in place in the affected area. Additionally, knowledge of the income of the area's residents, as suggested in the *special populations* factor, allows the agency to make some projections regarding the likelihood of insurance coverage, particularly special hazard insurance such as flood or earthquake insurance, which are potentially expensive additions to a homeowners policy. As one analyst has noted:

> The decision not to buy insurance for earthquakes, floods, hurricanes and other natural disasters isn't always conscious: some homeowners don't realize they're not already covered. But many others, faced with high premiums and policies with limited coverage, gamble that they won't need insurance help to rebuild after a disaster.64

Average Amount of Individual Assistance by State

The last factor FEMA considers in assessing IA needs is the *average amount of Individual Assistance by state.* FEMA has issued statistics on average losses but notes that the average numbers used are not a threshold (see *Table 2*). The agency does suggest that the "following averages may prove useful to states and voluntary agencies as they develop plans and programs to meet the needs of disaster victims." The inference is that the levels listed generally are what would be expected in damage to dwellings.

Table 2. Average Amount of Federal Assistance Per Disaster Based on Size of State, July 1995 - July 1999

Categories	Small States <2MillionPopulation	Medium States 2-10 Million Population	Large States >10 Million Population
Avg. population (1990 Census Data)	1,000,057	4,713,548	15,552,791
Number of disaster housing applications approved	1,507	2,747	4,679
Number of homes est. major damage/destroyed	173	582	801
Dollar amount of housing assistance	$2.8 million	$4.6 million	$9.5 million
Number of Individual and Family Grants (IFG) approved (Now known as ONA)	495	1,377	2,071
Dollar amount of IFG assistance (now ONA)	$1.1 million	$2.9 million	$4.6 million
Disaster housing/IFG (ONA) combined assistance	$3.9 million	$7.5 million	$ 14.1 million

Source: 44 CFR §206.48(b) (6).

The average numbers that follow are based on disasters that occurred between July of 1995 and July of 1999; the data are 10 years old and of questionable use today. The chart divides states into three categories: small states (under 2 million in population), medium states (2 to 10 million in population) and large states (over 10 million in population). The population amounts are based on the 1990 Census.

While FEMA's regulations stress that these are not thresholds, they are considered by agency officials in determining whether IA will be provided. Presumably states may consider that FEMA help could be forthcoming if damage reaches the state indicator levels. However, since the amounts have not been updated in ten years and are based on a 1990 census, it is difficult to determine the degree to which these numbers are considered. Given Congress's mandate in Section 320 of the Stafford Act, this cannot be an arithmetic formula that solely determines whether assistance is provided. But the presentation of loss indicators may guide states when considering whether to request assistance.

In addition to the issue of the data's currency, a larger and more compelling question is whether the numbers in the chart match up to the Preliminary Damage Assessments (PDAs) for those events. Absent an existing review of detailed information in PDA forms, it is not possible to determine the usefulness of the data in *Table 2*.[65] Still, given the level of experience in this field following literally thousands of disaster declarations over the last 30 years, it could be argued that the numbers of eligible households assisted may be reflective of the estimated damage upon which the decision for a disaster declaration was made.

Congressional Considerations for the Declaration Process

When Congress considers the mandated GAO reports and other commentary on the declaration process, there are some considerations that Members may wish to review when considering the current declaration process.

The Composition of Preliminary Damage Assessment Teams

One area of consideration is the composition of Preliminary Damage Assessment teams. Team members are involved throughout the process and include local officials guiding the team, state personnel who assist the governor in requesting assistance, and FEMA staff who work on the disaster if one is declared. While FEMA staff have the opportunity at several levels to refine the information the team gathers on damages and to ask additional questions, the process could be approached in other ways. Some FEMA staff have suggested the formation of several permanent teams that would have PDAs as their prime job task without continuing involvement in particular disasters. This could result in more consistent assessments with the bonus of added perspective of team members with exposure to various disasters in many regions of the country. Establishing independent and expert on-site groups to review a situation is a recommendation that the National Transportation Safety Board (NTSB) has been contemplating in their investigations of transportation accidents.[66]

The PDA is an important part of the declaration process. While it can be subject to challenge, it is also the assessment closest to the event. Given its vital role in the process the PDA deserves close attention to determine if these on-site assessments are accurately reflecting the character of an event and the likely eligible damages following a disaster event.

Updating and Revising Individual Assistance Averages

As previously noted, the averages that appear in FEMA's declaration process regulations for Individual Assistance (IA) to states are derived from experiences from July 1995 to July 1999.[67] These figures could be updated based on FEMA's more recent experience in delivering this type of assistance. Legislation introduced in the 111th Congress had a provision that directed FEMA to review the factors that are considered for an IA designation.[68]

While the averages are separated by state population size, there does not appear to be a threshold number equivalent to the PA figure of $1 million in eligible damage. In the case of IA help, a dollar figure may not be desirable; however, numbers of families and households affected, or a minimum number of homes with major damage, could potentially be a starting point in establishing a base IA threshold for consideration of requests.

Other Potential Disaster Indicators

The current per capita indicator based on state population according to the U.S. Census is clear, but some observers believe it lacks precision. For example, the earlier discussion on PA indicators pointed out the commonality between California and Nevada regarding earthquake risk as well as the growth of the states. Their population sizes are very different but the per capita indicator alone does not necessarily measure a state's fiscal capacity.

The GAO report in 2001 noted that per capita personal income measures do not take into account the taxable income a state may enjoy from businesses or corporations that may have considerable taxable profits. GAO did suggest a different indicator:

> We have previously reported that Total Taxable Resources (TTR), a measure developed by the U.S. Department of Treasury, is a better measure of state funding capacity in that it provides a more comprehensive measure of the resources that are potentially subject to state taxation.[69]

FEMA's response to this suggestion questioned whether the use of TTR would be in violation of the legislative prohibition against arithmetic formulas.[70] The TTR does have very specific information and also tracks growth within a state. However, this appears to make it a more accurate measurement. The degree of detail in the measurement would not necessarily make it the "sole" determinant of disaster aid. It could remain an indicator and one among several factors to be considered in the declaration process.

CONCLUSION

The disaster declaration process, though subject to inquiry, argument, hearings, studies and recommendations, has changed very little over time. It remains a process that can be observed and evaluated as it occurs in the area affected by the disaster, and grows opaque as it moves up through layers of FEMA and DHS management to the White House. By making PDA information available, FEMA has begun to lift the veil on the decisions that are made. Congress has demonstrated an interest in this process and has sought to understand its limits and its effectiveness. "From the Government Accountability Office in 1981 to Vice President Gore's National Performance Review in 1994," one student of the process noted, "countless policy critiques have called for more objective criteria for presidential disaster declarations."[71] But those calls can be muted when a disaster occurs in a particular place that has particular importance to actors in the process, whether in Congress or the executive branch.

While criteria have been established to create a relatively uniform and knowable process, these criteria may not be determinative of the most critical elements considered. An emphasis on victims of unexpected natural disaster events will likely always have a compelling influence on the disaster declaration process. But since disaster relief does have a political element, at many levels, precedent arguably can have value in improving the declaration process so that it can be applied broadly and fairly. But precedent can also be problematic when discussing events of very different size and impact.

During the last Congress, attention was directed to the nature of the disaster declaration process. The 112th Congress may choose to consider a broad review of the process that might include the consistency of FEMA's approach across the nation in making damage assessments, other potential indicators of state capabilities and capacities, and the currency of the factors it employs to evaluate those assessments of disaster damage and the state requests on which they are based.

End Notes

[1] 42 U.S.C. § 5122.

[2] Ibid.

[3] For additional discussion, see CRS Report RL34724, *Would an Influenza Pandemic Qualify as a Major Disaster Under the Stafford Act?*, by Edward C. Liu.

[4] For more information on disaster supplementals, see CRS Report RL33226, *Emergency Supplemental Appropriations Legislation for Disaster Assistance: Summary Data*, by Justin Murray and Bruce R. Lindsay. For a map of disaster declarations, see http://www.bakerprojects.com.fema/Maps/CONUS.pdf.

[5] Rutherford H. Platt, *Disasters and Democracy: The Politics of Extreme Natural Events*, (Washington, DC: Island Press, 1999), p. 66.

[6] 42 U.S.C. § 5170.

[7] 42 U.S.C. § 5191.

[8] Michael J. Trebilcock and Ronald J. Daniels, "Rationales and Instruments for Government Intervention in Natural Disasters" in Ronald J. Daniels, Donald F. Kettl, and Howard Kunreuther, eds., *On Risk and Disaster: Lessons from Hurricane Katrina*, (Philadelphia: University of Pennsylvania Press, 2006), p. 104. For additional discussion of the evolution of emergency management see Claire B. Rubin, ed. *Emergency Management: The American Experience 1900-2005* (Fairfax, VA: Public Entity Risk Institute, 2007).

[9] U.S. Department of Homeland Security, Federal Emergency Management Agency, Annual Report of FOIA Activity for FY2000, II. C., at http://www.fema.gov/doc/help/foiareportfy00.doc.

[10] U.S. Department of Homeland Security, Federal Emergency Management Agency, *Preliminary Damage Assessment Reports,* at http://www.fema.gov/rebuild/recover/pda-reports.shtm.

[11] Ibid.; and Platt, p. 266.

[12] U.S. Government Accountability Office, *Disaster Assistance: Improvements Needed in Disaster Declaration Criteria and Eligibility Assurance Procedures,* GAO-01-837, August, 2001, p. 2.

[13] For more information on emergency supplemental spending for disasters see CRS Report RL33226, *Emergency Supplemental Appropriations Legislation for Disaster Assistance: Summary Data*, by Justin Murray and Bruce R. Lindsay.

[14] U.S. Congress, House Committee on the Budget, *Budgetary Treatment of Emergencies,* p. 68, 105th Congress, 2nd. Sess., June 23, 1998 (Washington: GPO, 1998).

[15] For more information, see CRS Report RL33053, *Federal Stafford Act Disaster Assistance: Presidential Declarations, Eligible Activities, and Funding*, by Keith Bea.

[16] Ibid.; and Platt, p. 10.

[17] U.S. Congress, Senate Committee on Appropriations, Subcommittee on VA, HUD and Independent Agencies, *Departments of Veterans Affairs and Housing and Urban Development and Independent Agencies Appropriations for Fiscal Year 2002,* 107th Congress, 1st sess., p. 252, at http://www.access.gpo.gov/congress/senate.

[18] Opening Statement of Rep. Bennie Thompson, in U.S. Congress, House Committee on Homeland Security, *Disaster Declarations, Where is FEMA in a Time of Need?,* 110th Congress, 1st. sess., March 15, 2007.

[19] P.L. 110-28, 121 Stat. 155.

[20] U.S. Congress, House Committee on Appropriations, *Making Emergency Supplemental Appropriations for the Fiscal Year ending September 30, 2007 and other purposes, report to accompany H.R. 1591*, 110th Cong., 1st sess., March 20, 2007, H.Rept. 110-60, p. 211.

[21] Government Accountability Office, *FEMA Disaster Cost Estimates,* Feb. 2008, p. 4.

[22] Matt Korade, "FEMA's Success Hinges on Regional Personnel," *Congressional Quarterly,* March 16, 2010, at http://homeland.cq.com/hs/display/do?docid=3569657&sourcetype=31.

[23] P.L. 109-295, 689g, 120 Stat. 1453.

[24] For more information see CRS Report RL33729, *Federal Emergency Management Policy Changes After Hurricane Katrina: A Summary of Statutory Provisions*, coordinated by Keith Bea.

[25] U.S. Congress, House Subcommittee on Investigations and Oversight, *The Federal Emergency Management Agency's Proposed Disaster Relief Regulations,* 100th Cong., 1st sess. (Washington: GPO, 1987), pp. 3-5.

[26] 42 U.S.C. 5163.

[27] For more information, see CRS Report RL33090, *Robert T. Stafford Disaster Relief and Emergency Assistance Act: Legal Requirements for Federal and State Roles in Declarations of an Emergency or a Major Disaster*, by Elizabeth B. Bazan.

[28] 42 U.S.C. §5191.

[29] President Carter (Georgia), President Reagan (California), President Clinton (Arkansas), and President George W. Bush (Texas).

[30] Ibid.; and Platt, p. 20.

[31] CFR 44 § 206.38.

[32] CFR 44 § 206.33.

[33] Richard Sylves, Quick Response Report #86: *The Politics and Administration of Presidential Disaster Declarations: The California Floods of Winter 1995,* at http://www.colorado.edu/hazards/research/qr/qr86.html.

[34] 44 CFR §206.33(c).

[35] FEMA has 10 regional offices: Region 1 (Boston, MA), Region 2 (New York, NY), Region 3 (Philadelphia, PA), Region 4 (Atlanta, GA), Region 5 (Chicago, IL), Region 6 (Denton,

TX), Region 7 (Kansas City, MO), Region 8 (Denver, CO), Region 9 (Oakland, CA), and Region 10 (Bothell, WA).

[36] Sally Kestin and Megan O'Matz, "FEMA ruled on disaster before verifying Dade damage," *South Florida Sun-Sentinel*, at http://www.sun-sentinel.com/news/sfl-fema15may15,0, 6848867.story?coll=sfla-news-utilities.

[37] U.S. Department of Homeland Security, Federal Emergency Management Agency, "Joint State-Federal Preliminary Damage Assessment Teams Visit Flood-Damaged Communities," Release Number R10-06-047, November 13, 2006, at http://www.fema.gov /news/newsrelease.fema?id=31509.

[38] 44 CFR 206.33(d)

[39] Ibid.; and Sylves, p. 5.

[40] U.S. Department of Homeland Security, Federal Emergency Management Agency, *Presidential Declarations: What Does This Mean for Me?,* at http://www.fema.gov/ media/archives/2007/010807b.shtm.

[41] U.S. Department of Homeland Security, Federal Emergency Management Agency, "Notice of Adjustment of Statewide Per Capita Impact Indicator," 75 *Federal Register,* 62135, October 7, 2010.

[42] U.S. Census Bureau, *State & County Quick Facts,* at http://quickfacts.census.gov/gfd/states/ 06000.html.

[43] 42 U.S.C. §5170.

[44] 44 CFR § 206.48.

[45] U.S. Geological Survey, Earthquake Hazards Program, "Earthquake Density Maps for the United States", at http://earthquake.usgs.gov/regional/states/us_density.php.

[46] U.S. Department of Homeland Security, Federal Emergency Management Agency, "Notice of Adjustment of Countywide Per Capita Impact Indicator," 75 *Federal Register,* 62135, October 7, 2010.

[47] For information on public insurance coverage, see "Cost of Risk Survey" conducted by the Public Entity Risk Institute (PERI) at http://www.riskinstitute.org.

[48] 44 CFR §206.253. For additional information, see CRS Report RS22945, *Flood Insurance Requirements for Stafford Act Assistance*, by Edward C. Liu.

[49] 44 CFR §206.252.

[50] 42 U.S.C. § 5170c.

[51] For more information, see CRS Report RS22268, *Repairing and Reconstructing Disaster-Damaged Roads and Bridges: The Role of Federal-Aid Highway Assistance*, by Robert S. Kirk.

[52] For additional discussion of this law, see CRS Report RL34209, *Commercial Fishery Disaster Assistance*, by Harold F. Upton.

[53] For more information, see CRS Report RS21212, *Agricultural Disaster Assistance*, by Dennis A. Shields and Ralph M. Chite.

[54] 44 CFR §206.48(b)(1).

[55] P.L. 109-295, §689g, 120 Stat. 1453.

[56] 44 CFR §206.48(b)(2).

[57] 42 U.S.C. § 5174.

[58] Richard Sylves, Quick Response Report #86: *The Politics and Administration of Presidential Disaster Declarations: The California Floods of Winter 1995*, at http://www.colorado.edu/ hazards/research/qr/qr86.html

[59] 44 CFR §206.48 (b) (4).

[60] National Emergency Management Association (NEMA) 2006 Biennial Report: Organizations and Funding for State Emergency Management and Homeland Security, (Lexington, KY: Council of State Governments).

[61] For more information on state assistance programs, refer to the following for summary information on state statutory authorities: CRS Report RL32287, *Emergency Management and Homeland Security Statutory Authorities in the States, District of Columbia, and Insular Areas: A Summary*, by Keith Bea, L. Cheryl Runyon, and Kae M. Warnock.

[62] Ibid.; and Platt, p. 65.

[63] 42 U.S.C. §5155.

[64] Liz Pulliam Weston, "Do You Really Need Disaster Insurance?," *MSN Money,* at http://moneycentral.msn.com/content/Insurance/Insureyourhome/P59648.hsp?Printer.

[65] This statement remains accurate. However, since FEMA has begun to list PDA information on its website, it should be possible to begin to assess the efficacy of the factors employed to determine IA damage and declaration criteria.

[66] For more information, see CRS Report RL33474, *Reauthorization of the National Transportation Safety Board (NTSB)*, by Bart Elias.

[67] 44 CFR § 206.48.

[68] H.R. 3377, Title III, Sec.304.

[69] U.S. Government Accountability Office, *DISASTER ASSISTANCE: Improvement Needed in Disaster Declaration Criteria and Eligibility Assurance Procedures,* GAO-01-837, August 2001, p.11.

[70] Ibid., p. 49.

[71] Ibid.; and Platt, p. 285.

In: Federal Emergency Management
Editor: Lewis B. Adams
ISBN: 978-1-62618-217-2

Chapter 5

FEDERAL DISASTER RECOVERY PROGRAMS: BRIEF SUMMARIES*

Carolyn V. Torsell

SUMMARY

This report summarizes federal disaster assistance programs for use by Members of Congress and their staff as they address the needs of their states, communities, and constituents. A number of federal agencies provide assistance to individual victims, such as state, territorial, and local governments and non-governmental entities following a disaster. The federal forms of assistance include grants, loans, loan guarantees, temporary housing, and counseling.

The programs summarized in this report fall into two broad categories. First are programs for disaster situations; most of these programs are administered by the Federal Emergency Management Agency (FEMA) and the Small Business Administration (SBA). Second are general assistance programs that may be used in either disaster situations, or to meet regular service needs. Many federal agencies, including the Departments of Health and Human Services, Housing and Urban Development, and Justice, administer programs that may be included in the second category.

---

* This is an edited, reformatted and augmented version of a Congressional Research Service publication, CRS Report for Congress RL31734, from www.crs.gov, prepared for Members and Committees of Congress, dated August 10, 2012.

The programs in this report are primarily organized by recipient: individuals, state and local governments, or businesses. These programs address short-term needs such as food and shelter, and long-term needs such as repair of public utilities.

This report includes a list of relevant federal agency websites that provide information on disaster responses, updates on recovery efforts, and resources on federal assistance programs.

INTRODUCTION

Federal agencies provide a range of assistance to individual victims; state, territorial, and local governments; and non-governmental entities after major disasters, including certain terrorist attacks. Types of aid include, but are not limited to, technical assistance, loans and loan guarantees, grants, temporary housing, access to counseling professionals, and medical assistance. The full range of federal disaster relief assistance is available after the President issues a major disaster declaration pursuant to Stafford Act authority.[1] More limited aid is available under a Stafford Act emergency declaration[2] or a declaration issued by a department or agency head.[3]

This report identifies programs pertinent to the disaster recovery process. It provides brief descriptive information to help congressional offices determine which programs merit further consideration in the planning, organization, or implementation of recovery operations. Most of the programs listed here are authorized as assistance programs and listed in the *Catalog of Federal Domestic Assistance* (CFDA). The list does not include ad hoc assistance that some agencies provide in emergency or disaster situations.

PROGRAM SELECTION CRITERIA

Programs discussed in this report satisfy one or more of the following criteria:

- Congress expressly designated the program as a disaster relief or recovery program.
- The program is applicable to disaster situations, even if not specifically authorized for that purpose.

- The Federal Emergency Management Agency (FEMA) and other federal agencies have used the program as an emergency response resource.
- The program is potentially useful for addressing short-term and long-term recovery needs (e.g., assistance with processing survivor benefits or repair of public facilities).

Most of the programs listed in this report are specifically authorized for disaster situations. General assistance programs that may apply to disaster situations are described at the end of this report.

FEDERAL DISASTER RECOVERY PROGRAMS

Assistance for Individuals and Families

Individuals and Households Program

The Individuals and Households Program (IHP) is the primary vehicle for FEMA assistance to individuals after the President issues a major disaster declaration. Congress appropriates money for the IHP (and other aid authorized by the Robert T. Stafford Disaster Relief and Emergency Assistance Act) to the Disaster Relief Fund. Grants assist disaster victims whose needs cannot be met through other forms of assistance, such as insurance or other federal programs. Program funds have a wide range of eligible uses, including temporary housing; limited housing repair or replacement; and uninsured medical, dental, or other personal needs (generally referred to as Other Needs Assistance [ONA]). While the federal share of temporary housing assistance is 100%, there is a statutory matching requirement that states contribute 25% of ONA funds. Grants may not exceed $29,900 per individual or household, adjusted annually to reflect consumer price changes; IHP assistance is generally limited to a period of 18 months. (Also see "Physical Disaster Loans—Residential SBA Disaster Loans Available to Homeowners and Renters" below, for additional assistance for homeowners and renters.)

Agency: Federal Emergency Management Agency
Authority: 42 U.S.C. §§5174-5188
Regulation: 44 C.F.R. §§206.110-120
Phone: Office of Congressional Affairs, 202-646-4500

Website: http://www.fema.gov/public-assistance-local-state-tribal-and-non-profit/recovery-directorate/assistance-individuals-and
CFDA: 97.048-97.050

Disaster Unemployment Assistance

Disaster Unemployment Assistance (DUA) provides benefits to previously employed or self-employed individuals rendered jobless as a direct result of a major disaster, and who are not eligible for regular federal/state unemployment compensation (UC). In certain cases, individuals who have no work history or are unable to work may also be eligible for DUA benefits. DUA is federally funded through FEMA, but is administered by the Department of Labor and state UC agencies. Generally, individuals must apply for benefits within 30 days after the date the state announces availability of DUA benefits. When applicants have good cause, they may file claims after the 30-day deadline. This deadline may be extended; however, initial applications filed after the 26th week following the declaration date will not be considered. When a reasonable comparative earnings history can be constructed, DUA benefits are determined in a similar manner to regular state UC benefit rules. The minimum weekly DUA benefit is required to be half of the average weekly UC benefit for the state where the disaster occurred. DUA assistance is available to eligible individuals as long as the major disaster continues, but no longer than 26 weeks after the disaster declaration.

Agency: Department of Labor, Employment and Training Administration
Authority: 42 U.S.C. §5177
Regulation: 20 C.F.R. §625; 44 C.F.R. §206.141
Contact: See listing of contacts for state UC Agencies, http://www.servicelocator.org/OWSLinks.asp
Website: http://ows.doleta.gov/unemploy/disaster.asp
CFDA: 97.034

Dislocated Worker Activities

The dislocated worker program provides training and related assistance to persons who have lost their jobs and are unlikely to return to their current jobs or industries. Of the funds appropriated, 80% are allocated by formula grants to states and local entities; 20% are reserved by the Secretary of Labor for activities including national emergency grants to states or local entities. One type of national emergency grant is Disaster Relief Employment Assistance, under which funds can be made available to states to employ dislocated workers in temporary jobs involving recovery after a national emergency. An individual may generally earn up to $12,000, excluding the cost of fringe benefits, and be employed for up to six months or 1,040 hours, whichever is longer. An individual may earn the full $12,000 in less than six months or 1,040 hours by working in a higher skilled position for which a higher hourly wage is paid. There are no matching requirements for Workforce Investment Act programs.

Agency: Department of Labor, Employment and Training Administration
Authority: 20 U.S.C. §9201
Regulation: 20 C.F.R. §671
Contact: See listing of contacts for state Dislocated Worker/Rapid Response Coordinators at http://www.doleta.gov/layoff/rapid_coord.cfm
Website: http://www.doleta.gov/layoff/workers.cfm
CFDA: 17.260

Public Safety Officers' Benefits Program

This program provides benefits as lump-sum payments, currently $323,035.75 for eligible deaths occurring on or after October 1, 2011, to spouses and children of public safety officers who have been permanently disabled or have lost their lives while in the line of duty. The benefit may be adjusted each year to reflect the percentage of change in the Consumer Price Index (CPI). There is no matching requirement for this program.

Agency: Department of Justice, Bureau of Justice Assistance
Authority: 42 U.S.C. §3796
Regulation: 28 C.F.R. §32.31 *et seq.*
Phone: Office of Justice Programs, PSOB, 202-307-0635
Website: https://www.bja.gov/ProgramDetails.aspx?Program_ID=78
CFDA: 16.571

Public Safety Officers' Educational Assistance Program

This program provides an educational assistance allowance to eligible survivors of public safety officers whose death or total permanent disability are the direct and proximate result of an injury sustained in the line of duty. Funds may be used only to defray educational expenses, such as tuition, room and board, and books.

There is no matching requirement for this program.

Agency: Department of Justice, Bureau of Justice Assistance
Authority: 42 U.S.C. §3796d
Regulation: 28 CFR §32.31 to 32.37
Phone: Office of Justice Programs, PSOB, 202-307-0635
Website: https://www.bja.gov/ProgramDetails.aspx?Program_ID=78
CFDA: 16.615

Physical Disaster Loans—Residential SBA Disaster Loans Available to Homeowners and Renters

The majority of disaster loans provided by SBA (approximately 80%) are made available to individuals and households rather than businesses. SBA disaster assistance is provided in the form of loans, not grants, and therefore must be repaid to the federal government. Homeowners, renters, and personal property owners located in a declared disaster area (and in contiguous counties) may apply to the SBA for loans to help recover losses from the disaster. SBA's Home Disaster Loan Program falls into two categories: personal property loans and real property loans. These loans cover only uninsured or underinsured property and primary residences. Loan maturities may be up to 30 years.

Personal Property Loans

A personal property loan provides a creditworthy homeowner or renter with up to $40,000 to repair or replace personal property items such as furniture, clothing, or automobiles damaged or lost in a disaster. These loans cover only uninsured or underinsured property and primary residences and cannot be used to replace extraordinarily expensive or irreplaceable items such as antiques, recreational vehicles, or furs.

Real Property Loans

A creditworthy homeowner may apply for a "real property loan" of up to $200,000 to repair or restore the homeowner's primary residence to its pre-disaster condition. The loans may not be used to upgrade homes or build additions, unless upgrades or changes are required by city or county building codes. A real property loan may be increased by 20% for repairs to protect the damaged property from a similar disaster in the future.

Agency: Small Business Administration
Authority: 15 U.S.C. §636(b)
Regulation: 13 C.F.R. §§123.200-204

Phone: Office of Congressional and Legislative Affairs, 202-205-6700
Website: http://www.sba.gov/content/home-and-personal-property-loans
CFDA: 59.008

Cora Brown Fund

This unique fund directs payments to individuals and groups for disaster-related needs that have not been or will not be met by government agencies or other organizations. Annual obligations for the entire program are typically less than $225,000. There is no matching requirement for this program and no limitation on the time period in which assistance is available.

Agency: Federal Emergency Management Agency
Authority: 42 U.S.C. §5121 et seq.
Regulation: 44 C.F.R. §206.181
Phone: Office of Congressional Affairs, 202-646-4500
Website: http://www.gpo.gov/fdsys/pkg/CFR-2011-title44-vol1/pdf/CFR-2011-title44-vol1-sec206-181.pdf
CFDA: 97.031

Assistance for Victims of Crime

This program offers a variety of services to victims, including emergency travel assistance, assistance with obtaining information, and long-term mental health counseling. There is no matching requirement for this program.

Agency: Office of Justice Programs, Office for Victims of Crime, Department of Justice
Authority: 42 U.S.C. §10601
Regulation: no specific regulation
Phone: Victim Assistance Center, 1-800-331-0075
Website: http://www.ojp.usdoj.gov/ovc/
CFDA: 16.575, 16.576

Crisis Counseling

This program provides grants that enable states to offer immediate crisis counseling services, when required, to victims of a major federally declared disaster for the purpose of relieving mental health problems caused or aggravated by a major disaster or its aftermath. Assistance is short-term and community-oriented. Cost-share requirements are not imposed on this assistance. The regulations specify that program funding generally ends after nine months, but time extensions may be approved if requested by the state and approved by federal officials.

Agency: Federal Emergency Management
Agency Authority: 42 U.S.C. §5183
Regulation: 44 C.F.R. §206.171
Phone: Office of Congressional Affairs, 202-646-4500
Website: http://www.fema.gov/public-assistance-local-statedirectorate/crisis-counseling
CFDA: 97.032

Disaster Legal Services

These services offer free legal assistance to individuals affected by a major federal disaster. Types of assistance typically include help with insurance claims, preparing powers of attorney, help with guardianships, and preparing new wills and other lost legal documents. Neither the statute nor the regulations establish cost-share requirements or time limitations for assistance.

Agency: Federal Emergency Management Agency
Authority: 42 U.S.C. §5182
Regulation: 44 C.F.R. §206.164
Phone: Office of Congressional Affairs, 202-646-4500
Website: http://www.fema.gov/public-assistance-local-state-tribal-and-non-profit/recovery-directorate/disaster-legal-services
CFDA: 97.033

Tax Relief

Individuals and businesses located in areas included in major disaster declarations are allowed extra time (four years for homes and two years for business property) to replace property due to involuntary conversion (e.g., due to destruction from wind or floods, theft, or property ordered to be demolished). Casualty losses can also be deducted against income in the prior tax year, and certain disaster relief grants are not taxable. These losses also include property loss due to government ordered demolition because the property is unsafe. Disaster relief payments for certain expenses are not subject to tax. The Internal Revenue Service also has the authority to extend tax filing deadlines. Special provisions were enacted for the Katrina disaster area.[4] That law eliminates floors for the casualty loss deduction, extends the replacement period for involuntary conversions to five years, allows penalty free withdrawals from IRAs and retirement plans, allows earned income and child credits based on prior year incomes, allows tax credits for retaining and hiring residents, and allows exclusions for cancellation of indebtedness.

Agency: Internal Revenue Service
Authority: Internal Revenue Code, Title 26 *United States Code*
Regulation: No specific regulation
Phone: Congressional Liaison, 202-622-9590
Website: http://www.irs.gov/newsroom/article/0,,id=108362,00.html

Assistance for State, Territorial, and Local Governments

Public Assistance Grants

Authorized by the Stafford Act, this is FEMA's primary assistance program for state and local governments. Public Assistance (PA) grants may be used to repair, replace, or restore disaster-damaged, publicly owned facilities and the facilities of certain private nonprofit organizations that perform a governmental function. This program has a wide range of eligible activities, including removal of debris, repair of roads and bridges, and repair of public buildings and water control facilities. There is generally a 25% matching requirement, but the statute authorizes the President to waive part or all of the required match. The regulations specify that federal officials must obligate funds for approved applications within 45 days of the approval of applications for assistance (referred to as "Project Worksheets"). (Also see "Physical Disaster Loans— Businesses," below, for assistance for nonprofit organizations.)

Agency: Federal Emergency Management Agency
Authority: 42 U.S.C. 5172 §(a)(1)(A)
Regulation: 44 CFR §206, subparts G, H, I
Phone: Office of Congressional Affairs, 202-646-4500
Website: http://www.fema.gov/public-assistance-local-state-tribal-and-non-profit
CFDA: 97.036

Hazard Mitigation Grants

The Hazard Mitigation Grant Program (HMGP) provides grants to states to implement mitigation measures after recovery from a disaster and to provide funding for previously identified mitigation measures to benefit the area. There is a matching requirement of 25%. The Department of Homeland Security Appropriations Act of 2007 adjusts the percentage amounts for HMGP awards by establishing a scale that authorizes a higher percentage for hazard mitigation grants (15% of the total Stafford Act assistance in a state) for a major disaster in which no more than $2 billion is provided, to 10% for assistance that ranges from more than $2 billion to $10 billion, and to 7.5% for a major disaster that involves Stafford Act assistance from more than $10 billion to $35.3 billion.[5]

Agency: Federal Emergency Management Agency
Authority: 42 U.S.C. §5170c
Regulation: 44 C.F.R. §§206.430 to 440
Phone: Office of Congressional Affairs, 202-646-4500
Website: http://www.fema.gov/hazard-mitigation-grant-program-hmgp/what-hazard-mitigation-grant-program
CFDA: 97.039

Pre-Disaster Mitigation Grants

This program provides grants and technical assistance to states, territories, and local communities for cost-effective hazard mitigation activities that complement a comprehensive hazard mitigation program, and reduce injuries, loss of life, and damage and destruction of property. A minimum of $575,000 or 1.0% of appropriated funds is provided to a state or local government, with assistance capped at 15% of appropriated funds. Federal funds generally comprise 75% of the cost of approved mitigation projects, except for small impoverished communities that may receive up to 90% of the cost.

Agency: Federal Emergency Management Agency
Authority: 42 U.S.C. §5133
Regulation: 44 C.F.R. §201
Phone: Office of Congressional Affairs, 202-646-4500
Website: http://www.fema.gov/pre-disaster-mitigation-grant-program
CFDA: 97.047

Community Disaster Loans

This program provides loans to local governments that have suffered substantial loss of tax and other revenue in areas included in a major disaster declaration. The funds can only be used to maintain existing functions of a municipal operating character. The local government must demonstrate a need for financial assistance. Typically, the loan may not exceed 25% of the local government's annual operating budget for the fiscal year of the disaster. The limit is 50% if the local government lost 75% or more of its tax and other revenue as the result of a major disaster. A loan may not exceed $5 million. There is no matching requirement. The statute does not impose time limitations on the assistance, but the normal term of a loan is five years. The statute provides that the repayment requirement is cancelled if local government revenues are not sufficient to meet operations expenses during three years after a disaster. The governor must request the program specifically and funds must be appropriated into the CDL account by Congress.

Agency: Federal Emergency Management Agency
Authority: 42 U.S.C. §5184
Regulation: 44 C.F.R. §§206.360 to 378
Phone: Office of Congressional Affairs, 202-646-4500
Website: http://www.fema.gov/community-disaster-loan-program
CFDA: 97.030

Reimbursement for Firefighting on Federal Property

This program directs payments to fire departments that fight fires on federal property. Applicants may be reimbursed for direct expenses and direct losses (those not considered normal operating expenses) incurred in firefighting. Cost share and time limitations are not imposed in the statute or regulations.

Agency: Federal Emergency Management Agency
Authority: 15 U.S.C. §2201 to 2234
Regulation: 44 C.F.R. §§151 to151.23
Phone: Office of Congressional Affairs, 202-646-4500
Website: http://www.usfa.dhs.gov/grants/rfff/
CFDA: 97.016

Fire Management Assistance Grant Program

This program provides grants to state and local governments to aid states and their communities with the mitigation, management, and control of fires burning on publicly or privately owned forests or grasslands. The federal government provides 75% of the costs associated with fire management projects, but funding is limited to calculations of the "fire cost threshold" for each state.[6] No time limitation is applied to the program.

Agency: Federal Emergency Management Agency
Authority: 42 U.S.C. §5187
Regulation: 44 C.F.R. §§204.1 to 204.64
Phone: Office of Congressional Affairs, 202-646-4500
Website: http://www.fema.gov/fire-management-assistance-grants-program-details
CFDA: 97.046

Assistance for Small Businesses and Nonprofit Organizations

Economic Injury Disaster Loans

This program assists small businesses suffering economic injury as a result of disasters by offering loans and loan guarantees. Businesses must be located in disaster areas declared by the President, the Small Business Administration, or the Secretary of Agriculture. There is no matching requirement in this program. The maximum loan amount is $2.0 million. Loans may be up to 30 years. The application period is announced at the time of the disaster declaration.

Agency: Small Business Administration
Authority: 15 U.S.C. §636(b)
Regulation: 13 C.F.R. §§123.300 to 303
Phone: Office of Congressional Affairs, 202-205-6700
Website: http://www.sba.gov/content/economic-injury-disaster-loans

Physical Disaster Loans—Businesses

This program provides loans to small and large businesses in declared disaster areas for uninsured physical damage and losses. The maximum loan amount is $2.0 million. Loans may be for up to 30 years. There is no matching requirement in this program.

Agency: Small Business Administration
Authority: 15 U.S.C. §636(b)
Regulation: 13 C.F.R. §§123.200 to 204
Phone: Office of Congressional Affairs, 202-205-6700
*Website:*http://www.sba.gov/content/business-physical-disaster-loans/
CFDA: 59.008

Emergency Loans for Farms

When a county has been declared a disaster area by either the President or the Secretary of Agriculture, agricultural producers in that county may become eligible for low-interest emergency disaster (EM) loans available through the U.S. Department of Agriculture's Farm Service Agency. Producers in counties that are contiguous to a county with a disaster designation also become eligible for an EM loan. EM loan funds may be used to help eligible farmers, ranchers, and aquaculture producers recover from production losses (when the producer suffers a significant loss of an annual crop) or from physical losses (such as repairing or replacing damaged or destroyed structures or equipment, or for the replanting of permanent crops such as orchards). A qualified applicant can then borrow up to 100% of actual production or physical losses (not to exceed $500,000) at a below-market interest rate (which is currently 3.75%).

Agency: Department of Agriculture, Farm Service Agency
Authority: 7 U.S.C. §1961
Regulation: 7 C.F.R. §§1945.1 to1945.45
Phone: Legislative Liaison Staff, 202-720-3865
Website: http://www.fsa.usda.gov/FSA/webapp?area=home&subject=fmlp&topic=efl
CFDA: 10.404 and others

National Flood Insurance Program

Since 1968, the federal government has pursued a comprehensive flood risk management strategy designed to (1) identify and map flood-prone communities across the country (flood hazard mapping); (2) encourage property owners in NFIP participating communities to purchase insurance as a protection against flood losses (flood insurance); and (3) require communities in designated flood risk zones to adopt and enforce approved floodplain management ordinance to reduce future flood risk to new construction in regulated floodplains (floodplain management). The Federal Insurance and Mitigation Administration (FIMA), a part of the FEMA in the U.S. Department of Homeland Security (DHS), manages the NFIP.

Agency: Federal Emergency Management Agency
Authority: 42 U.S.C. 4001 et seq.
Regulation: 44 CFR §59.1–§82.21
Phone: Office of Congressional Affairs, 202-646-4500
Website: http://www.fema.gov/national-flood-insurance-program
CFDA: 97.022

General Assistance Programs

In addition to programs described above, which provide targeted assistance to individuals, states, territories, local governments, and businesses specifically affected by disasters, other general assistance programs may be useful to communities in disaster situations. For example, individuals who lose income, employment, or health insurance may become eligible for programs that are not specifically intended as disaster relief, such as cash assistance under the Temporary Assistance for Needy Families (TANF) program, job training under the Workforce Investment Act, Medicaid, or the State Children's Health Insurance Program (S-CHIP). Likewise, state or local officials have the discretion to use funds under programs such as the Social Services Block Grant or Community Development Block Grant to meet disaster-related needs, even though these programs were not established specifically for such purposes. Other agencies may offer assistance to state and local governments, including the Economic Development Administration and the Army Corps of Engineers. For businesses, however, only the disaster programs administered by the Small Business Administration are generally applicable. There are numerous federal programs that potentially offer disaster relief. However, specific eligibility criteria or other program rules might make it less likely that they would actually be used. Moreover, available funds might already be obligated for ongoing program activities. To the extent that federal agencies have discretion in the administration of programs, some agencies may choose to adapt these non-targeted programs for use in disaster situations. Also, Congress may choose to provide additional funds through emergency supplemental appropriations for certain general assistance programs, specifically for use after a disaster.

CRS analysts and program specialists can help provide information regarding general assistance programs that might be relevant to a given disaster situation. CRS appropriations reports may have information on disaster assistance within particular federal agencies.[7] These reports also contain a list of key policy staff at CRS, and identify the staff by their program area and agency expertise.

Federal Agency Websites

Note: Because of limited information available on the Internet, congressional users of this report are encouraged to contact the appropriate

CRS program analysts or department or agency program officers for more complete, timely information.

http://www.USA.gov/

Many federal agencies have established websites specifically to respond to disasters. Some agencies maintain websites with comprehensive information about their disaster assistance programs, whereas other agencies supply only limited information; most list contact phone numbers. An *A-Z Index of U.S. Government Departments and Agencies* is available at the website above.

FEMA Website

http://www.fema.gov

From its website, FEMA offers regular updates on recovery efforts in areas under a major disaster declaration. Information on a specific disaster may include a listing of declared counties and contact information for local residents.

Disaster Assistance.gov

http://www.disasterassistance.gov/

This website provides information on how to obtain help from the U.S. government before, during, and after a disaster. DisasterAssistance.gov also provides news, information, and resources to prepare for a disaster, to use during an emergency, and to recover afterward.

Catalog of Federal Domestic Assistance

http://www.cfda.gov

The CFDA describes more than 2,200 federal assistance programs. CFDA program numbers are given in each program summary included in this report. For current appropriations and additional information users should contact departments and agencies.

End Notes

[1] 42 U.S.C. §5170. For more information on the Stafford Act, see CRS Report RL33053, Federal Stafford Act Disaster Assistance: Presidential Declarations, Eligible Activities, and Funding, by Francis X. McCarthy.

[2] 42 U.S.C. §5191.

[3] The Administrators of the Small Business Administration, Farm Services Agency, and the Secretaries of Agriculture and Transportation, within statutory limitations, may issue disaster declarations for purposes of providing assistance solely from their agencies. Citations to these authorities are presented in the relevant program summaries.

[4] Katrina Emergency Tax Relief Act of 2005, P.L. 109-73, 119 Stat. 2016 (2005).

[5] Department of Homeland Security Appropriations Act, 2007, P.L. 109-295, §684, 120 Stat. 1355, 1447 (2006) (amended §404 of the Stafford Act).

[6] 44 C.F.R. §204.51.

[7] See CRS website at http://www.crs.gov/Pages/clis.aspx?cliid=73&preview=False for CRS appropriations reports.

INDEX

#

9/11, 67
9/11 Commission, 67

A

abuse, 18
access, 44, 48, 76, 78, 129, 141, 146
accessibility, 76
accommodation, 75
accountability, 18, 75
adaptation, 4
adjustment, 88
administrators, x, 72, 101
age, 132
agricultural producers, 22, 157
agriculture, 7, 46, 50
Air Force, 21, 33
ambient air, 19
American Recovery and Reinvestment Act, 80, 101
American Recovery and Reinvestment Act of 2009, 80, 101
American Red Cross, 125
Americans with Disabilities Act, 103
appointees, 123
appropriations, viii, x, xi, 1, 13, 17, 25, 26, 31, 32, 71, 72, 74, 80, 84, 89, 90, 93, 94, 99, 102, 113, 115, 118, 158, 159, 160
Appropriations Act, 30, 107, 154, 160
Appropriations Committee, 119
aquaculture, 157
arbitration, x, 72, 80, 90, 99, 100, 101, 102, 109, 110
architect, 44, 48
arithmetic, 122, 127, 137, 139
assessment, 10, 47, 57, 66, 81, 82, 88, 118, 123, 125, 129, 130, 133, 134, 138
assets, 6, 23, 30, 32, 67, 68, 110, 114
Attorney General, 47, 50, 56, 65
audit, 18
authorities, vii, 1, 3, 9, 14, 16, 19, 20, 21, 22, 23, 24, 27, 32, 33, 34, 40, 45, 48, 49, 53, 55, 56, 57, 60, 62, 67, 75, 83, 88, 95, 105, 116, 117, 127, 143, 160
automobiles, 150
autonomy, 9
awareness, 44

B

background information, 106
bacteria, 130
base, 52, 80, 87, 106, 134, 138
benefits, x, 22, 23, 72, 80, 96, 106, 123, 135, 147, 148, 149

blueprint, 48
Border Patrol, 67
border security, 7, 46, 50
borrowers, 25
budget deficit, 115
building code, 5, 80, 150
Bureau of Justice Assistance, 149, 150
businesses, xii, 29, 31, 79, 94, 95, 124, 126, 139, 146, 150, 152, 156, 158

C

cash, 158
Catalog of Federal Domestic Assistance, 146, 159
catalyst, 130
catastrophes, x, xi, 71, 72, 73, 92, 97, 108, 113, 130, 133
catastrophic failure, 26
Census, 128, 136, 137, 138, 142
CERCLA, 19, 32
CFR, 15, 31, 100, 103, 104, 105, 106, 107, 110, 117, 136, 141, 142, 143, 150, 151, 153, 157
chain of command, 23
challenges, ix, 38, 41, 42, 54, 126
chemical, 33
chemicals, 27
Chicago, 141
children, 149
cities, 85
citizens, 41
City, 31, 93, 94, 142
civil action, 27
civil disorder, 21
Clean Air Act, 28
Clean Water Act, 19, 32
cleanup, 27, 28, 35
Clinton Administration, 79, 89
clothing, 150
CNN, 127
Coast Guard, 15, 20, 32, 33, 67
commercial, 15, 130
communication, 4
communication systems, 4
communities, x, xi, xii, 25, 34, 41, 46, 51, 63, 72, 74, 81, 90, 91, 93, 114, 115, 119, 121, 131, 132, 134, 135, 145, 154, 155, 157, 158
community, 12, 22, 25, 27, 28, 34, 43, 46, 51, 52, 57, 61, 74, 77, 78, 85, 91, 93, 102, 108, 124, 132, 133, 134, 151, 155
community service, 46, 52
compensation, 85, 148
compilation, 109
complement, 154
compliance, 35, 62, 63
complications, 126
composition, 137
Comprehensive Environmental Response, Compensation, and Liability Act, 32
conference, 94, 108
conflict, 58, 62
conformity, 62
congress, 141
congressional offices, 146
consent, 21
constituents, xii, 145
Constitution, 9, 34, 105, 116, 123
construction, 5, 24, 48, 53, 58, 84, 117, 157
Consumer Price Index, 149
contaminant, 27
contamination, 20, 92
contingency, 133
convention, 65
cooperation, 41, 125
cooperative agreements, 32, 34
coordination, viii, 2, 4, 20, 26, 41, 49, 56, 62, 86, 95
cost, viii, 2, 3, 4, 13, 17, 31, 38, 59, 60, 77, 79, 80, 81, 86, 88, 89, 90, 92, 101, 102, 103, 106, 108, 116, 118, 119, 120, 123, 126, 127, 129, 149, 152, 154, 155
cost of living, 103
cost-benefit analysis, 129

counseling, xii, 11, 85, 126, 131, 145, 146, 151, 152
counterterrorism, 45, 50, 56
CPI, 149
critical infrastructure, 46, 50, 51, 56, 57, 65, 66, 68
crop, 22, 33, 157
crop insurance, 22, 33
crops, 157
currency, 115, 137, 140
curricula, 8
customers, 93
cybersecurity, 46, 50

D

damages, 6, 11, 29, 31, 82, 83, 84, 105, 110, 126, 128, 129, 131, 135, 137, 138
danger, 32
deaths, 82, 149
decision makers, xi, 114
decision-making process, 132
deduction, 152
deficit, 96
degradation, 29
denial, 110, 125
Department of Agriculture, viii, 2, 3, 22, 157
Department of Commerce, 69
Department of Defense, viii, 2, 3, 22, 23, 24, 33, 34, 41, 56, 67, 81, 84, 104
Department of Energy, 28
Department of Health and Human Services, viii, 2, 3, 15, 16, 24, 32
Department of Homeland Security, x, 15, 18, 30, 32, 34, 35, 62, 66, 67, 68, 69, 71, 73, 99, 105, 106, 107, 110, 124, 140, 142, 154, 157, 160
Department of Justice, 40, 149, 150, 151
Department of Labor, 74, 148, 149
Department of the Interior, 35, 76
Department of Transportation, viii, 2, 3, 25, 32
deposits, 76
depression, 108
depth, 3
destruction, 17, 28, 79, 81, 89, 90, 93, 98, 152, 154
detection, 56
DHS, x, 17, 18, 31, 56, 59, 62, 66, 69, 71, 72, 73, 74, 80, 81, 124, 139, 157
direct payment, 22
directives, 33, 39, 41, 65, 67
disability, 75, 149
disappointment, 124
disaster area, 77, 150, 152, 156, 157
disaster assistance, x, xii, 9, 10, 12, 21, 22, 25, 31, 35, 59, 71, 72, 73, 75, 80, 84, 85, 86, 90, 96, 102, 115, 123, 130, 133, 145, 150, 158, 159
disaster relief, viii, 1, 10, 17, 26, 31, 86, 89, 92, 93, 95, 97, 100, 117, 118, 140, 146, 152, 158
Disaster Relief Fund, x, 13, 31, 71, 72, 74, 80, 89, 100, 102, 107, 115, 118, 120, 124, 147
discrimination, 75
dispersion, 131
disposition, 79, 99
dissatisfaction, 100
distribution, x, 12, 28, 71, 84, 85, 86, 93, 106
District of Columbia, 73, 102, 143
DOI, 35
DOT, 25, 26
draft, 7, 30, 43, 47, 58, 59, 60, 64, 122
drawing, 44, 48
DRF, x, 13, 14, 16, 31, 32, 71, 74, 89, 90, 99, 102, 115, 120, 124
drinking water, 27
drought, 10, 24, 34, 82, 130

E

earnings, 148
earthquakes, 5, 57, 58, 126, 129, 135

economic damage, 38
economic development, 46, 52
economic losses, 29
economic policy, 43
education, 61, 77
elected leaders, 122
electricity, 93
eligibility criteria, x, 72, 93, 158
Emergency Assistance, ix, xi, 9, 42, 71, 72, 102, 113, 114, 141, 147
emergency communications, 85
emergency declarations, x, 10, 12, 31, 71, 80, 98, 106, 114
emergency management, vii, viii, 1, 2, 3, 4, 5, 7, 8, 9, 17, 18, 19, 39, 41, 42, 53, 75, 79, 91, 95, 102, 109, 140
emergency management functions, vii, 1, 3, 4, 5
emergency preparedness, 17, 73, 95
Emergency Relief Program, 26, 34
emergency response, 6, 19, 20, 21, 24, 26, 27, 75, 147
employees, x, 72, 80, 81, 87, 106, 125
employment, 87, 91, 158
enemies, 42
energy, 28, 79
enforcement, 34
engineering, 48
environment, 6, 19, 20, 27, 46, 51, 56
environmental contamination, 20
environmental impact, 115
Environmental Protection Agency, viii, 2, 3, 15, 20, 27, 34, 67
EPA, 15, 20, 27, 28, 32, 35
equipment, 12, 21, 23, 28, 46, 53, 74, 75, 78, 81, 106, 157
equity, xi, 114, 117
evacuation, 26, 75, 81
evolution, 39, 40, 44, 64, 102, 140
execution, 6, 45, 53, 116, 120, 123
executive branch, 13, 40, 117, 139
Executive Order, 32, 39, 40, 65, 102
executive orders, 20, 32, 67, 73
exercise, 13, 61, 74, 105, 116, 123, 124, 125
exercise programs, 61
expenditures, 18, 92, 96, 97, 98, 109, 116, 118, 124
expertise, 55, 67, 110, 158
exposure, 138

F

FAA, 35
fairness, xi, 114
families, x, xi, 6, 18, 71, 74, 83, 96, 98, 113, 114, 115, 123, 131, 132, 138
family members, 16
farmers, 157
farmland, 22
federal agencies, viii, xi, xii, 2, 3, 7, 13, 14, 15, 18, 20, 21, 27, 41, 48, 55, 56, 64, 67, 72, 76, 86, 90, 91, 102, 104, 106, 113, 145, 147, 158, 159
federal agency, viii, xii, 1, 16, 23, 27, 74, 84, 90, 135, 146
federal aid, x, 25, 71, 73, 77, 82, 86, 88, 97, 98, 121, 124, 130
federal assistance, vii, viii, x, xii, 2, 3, 9, 10, 11, 12, 16, 33, 39, 61, 72, 74, 75, 77, 78, 84, 86, 88, 89, 92, 93, 95, 96, 105, 108, 115, 119, 124, 127, 128, 131, 146, 159
federal authorities, 49, 74
Federal Bureau of Investigation, 56, 67
federal emergency management, viii, 2, 3, 4, 7, 17, 18, 79, 95, 102, 109
Federal Emergency Management Agency, viii, x, xii, 1, 2, 3, 10, 15, 30, 32, 52, 53, 65, 66, 67, 68, 71, 106, 107, 108, 110, 115, 140, 141, 142, 145, 147, 151, 152, 153, 154, 155, 156, 157
federal facilities, 32
federal funds, 118
Federal funds, 154

federal government, vii, xi, 1, 3, 4, 5, 7, 8, 9, 13, 16, 19, 20, 25, 26, 40, 49, 53, 54, 59, 60, 76, 82, 88, 90, 91, 114, 119, 134, 150, 155, 157
Federal Highway Administration, 26, 130
federal law, viii, 2, 3, 19, 21, 34, 104
Federal Register, 59, 65, 69, 103, 107, 110, 142
federal regulations, 91
federal roles, ix, 38, 40
federal-aid highway system, 26
feedstocks, 33
FHWA, 26, 130
financial, 17, 47, 57, 63, 76, 82, 83, 84, 92, 118, 120, 135, 155
financial incentives, 57
financial system, 47
fire suppression, vii, viii, 2, 73
fires, 3, 4, 5, 28, 81, 155
first responders, 20, 39
fishing, 130
fixation, 42
flexibility, 25, 42
flooding, 6, 24, 58, 94
floods, 57, 103, 119, 135, 152
food, xii, 6, 7, 12, 23, 46, 50, 85, 86, 106, 146
food security, 7
force, 82, 96, 106, 108
formation, 138
formula, 12, 13, 25, 106, 122, 127, 137, 148
foundations, viii, 2, 3
fraud, 18
fringe benefits, 149
funding, viii, x, 2, 5, 13, 14, 15, 25, 26, 35, 60, 72, 74, 86, 88, 89, 90, 93, 100, 103, 107, 109, 118, 120, 128, 130, 134, 139, 151, 153, 155
funds, vii, 1, 11, 13, 20, 25, 26, 34, 78, 79, 83, 84, 88, 89, 90, 91, 93, 94, 99, 102, 105, 108, 110, 115, 118, 120, 130, 133, 147, 148, 153, 154, 157, 158
fusion, 56

G

GAO, 97, 109, 118, 120, 122, 137, 139, 141, 143
General Accounting Office, 104, 109
Georgia, 119, 141
God, 117
good deed, 129, 134
governance, 9, 96
government revenues, 155
governments, xii, 12, 17, 76, 77, 78, 89, 129, 158
governor, xi, 9, 10, 12, 22, 31, 33, 34, 81, 82, 114, 118, 123, 124, 125, 126, 128, 130, 132, 133, 137, 155
grant programs, 57, 61, 108
grants, xii, 5, 11, 12, 22, 24, 25, 28, 57, 65, 74, 76, 77, 78, 79, 80, 81, 84, 85, 86, 87, 88, 89, 90, 91, 92, 94, 95, 101, 106, 107, 108, 145, 146, 148, 150, 151, 152, 153, 154, 155, 156
grasslands, 81, 155
groundwater, 19, 20
grouping, 5
growth, 117, 139
guidance, ix, 27, 38, 40, 41, 49, 53, 55, 60, 61, 63, 100, 125
guidelines, 18, 75, 83
Gulf Coast, viii, x, 2, 13, 26, 72, 90, 99

H

Haiti, 64
harbors, 20
hazardous materials, 21, 28, 133
Hazardous Substance Superfund Trust Fund, 20, 33
hazardous substances, 19, 20, 21, 27

hazards, vii, ix, 1, 3, 4, 5, 17, 24, 30, 38, 42, 43, 44, 47, 48, 55, 57, 58, 104, 141, 142
health, x, 9, 15, 24, 27, 46, 52, 57, 72, 77, 79, 80, 91, 93, 94, 104, 158
Health and Human Services, xii, 145
health care, 27, 93, 94
health insurance, 158
HHS, 24, 34, 69
high winds, 5
highways, 26
hiring, 153
history, 67, 95, 102, 117, 121, 133, 148
home ownership, 133
homeland security, 7, 39, 41, 53, 56, 65
Homeland Security Act, 17, 18, 102, 103
homeowners, 29, 79, 125, 126, 135, 147
homes, 6, 29, 76, 82, 83, 85, 126, 136, 138, 150, 152
host, 42
hourly wage, 149
House, 32, 52, 66, 94, 104, 107, 108, 109, 111, 119, 121, 141
House Committee on Transportation and Infrastructure, 104, 107
housing, x, 6, 22, 25, 46, 52, 66, 68, 72, 75, 76, 79, 82, 83, 85, 87, 90, 92, 98, 101, 103, 107, 131, 136, 147
Housing and Urban Development, viii, xii, 2, 3, 25, 59, 93, 141, 145
HUD, 25, 34, 59, 69, 93, 141
human, vii, 1, 3, 6, 7, 8, 23, 24, 25, 27, 38, 43, 46, 51, 57, 81, 94
human health, 27, 81
human-caused hazards, vii, 1, 3
Hurricane Andrew, 96, 104, 119
Hurricane Katrina, x, 34, 35, 72, 73, 75, 91, 100, 114, 140, 141
hurricanes, viii, 2, 28, 90, 100, 104, 130, 135
hypothesis, 125

I

ICS, 4, 30
ID, 69, 149, 150
identification, 126
immigration, 46, 50, 56
impairments, 22
improvements, xi, 114
inauguration, 73
income, 33, 79, 80, 85, 98, 122, 133, 135, 139, 152, 158
individuals, x, xi, xii, 6, 12, 13, 18, 21, 29, 44, 71, 74, 75, 79, 83, 84, 85, 86, 90, 98, 101, 109, 113, 114, 115, 123, 124, 131, 135, 146, 147, 148, 150, 151, 152, 158
industries, 128, 148
industry, 130
inflation, 80
information sharing, 7, 45, 50, 56
infrastructure, x, xi, 12, 14, 26, 27, 31, 35, 46, 52, 56, 57, 67, 72, 79, 83, 84, 90, 93, 94, 113, 114, 122, 127, 128, 129, 131
injuries, 17, 82, 89, 132, 154
injury, 15, 29, 149, 156
inspections, 24
institutions, 91, 93, 94
integration, 53, 95
intelligence, 6, 7, 47, 52, 55, 56, 67
Intelligence Reform and Terrorism Prevention Act, 67
interagency coordination, 66
interference, 34
Internal Revenue Service, 152, 153
investments, 129
Iowa, 103
issues, ix, 3, 10, 38, 40, 56, 60, 83, 90, 105, 109, 146, 147
iteration, 58

J

Japan, 64
job training, 158
jobless, 148
jurisdiction, 17, 19, 32, 87, 89

L

landscape, 117
law enforcement, 21, 52, 56, 67
laws, 9, 17, 20, 29, 34, 52, 64, 65, 80, 105, 116, 123
lead, viii, 1, 20, 27, 44, 49, 55, 57, 59, 62, 92, 96, 97, 120, 124
leadership, 9, 17, 46, 50, 62, 73
learning, 61
legal protection, 22
legislation, x, xi, 17, 31, 60, 67, 72, 74, 79, 80, 84, 89, 91, 92, 93, 94, 95, 98, 99, 100, 111, 113, 114, 118, 122
levees, 6, 22, 24, 58
light, 96
livestock, 15
loan guarantees, xii, 145, 146, 156
loans, xii, 15, 22, 29, 35, 74, 77, 85, 90, 91, 92, 95, 96, 108, 133, 145, 146, 150, 151, 154, 156, 157
Louisiana, 99, 100, 110
lying, 6

M

magnitude, 10, 12, 96, 115, 116, 124
major disaster declaration, xi, 10, 13, 14, 16, 28, 73, 79, 81, 82, 83, 84, 85, 86, 88, 90, 92, 94, 97, 98, 105, 113, 114, 115, 116, 123, 130, 146, 147, 152, 154, 159
major issues, 62
majority, 9, 13, 21, 22, 26, 29, 33, 53, 54, 57, 58, 100, 118, 150
man, ix, 37, 42, 115
management, vii, viii, 1, 2, 3, 4, 5, 7, 8, 12, 18, 28, 34, 58, 72, 73, 75, 76, 80, 81, 89, 108, 120, 134, 139, 155, 156, 157
man-made disasters, 42
mapping, 6, 157
Marine Corps, 33
maritime security, 46, 50
mass, 12, 26, 66, 68, 74, 85, 115
mass care, 12, 66, 68, 74
mass media, 115
materials, 5, 117
matter, 55, 101
means tested, 98
measurement, 139
measurements, 82, 128
media, 66, 124, 125, 127, 132, 142
Medicaid, 158
medical, 6, 11, 12, 16, 23, 24, 34, 66, 68, 75, 85, 87, 91, 94, 106, 132, 146, 147
medical assistance, 146
medical care, 16, 23, 106
medication, 6
medicine, 12, 86
mental health, 85, 151
metaphor, 48, 53, 55
methodology, ix, 38, 47, 66
military, 23, 34
militia, 21
mission, ix, 7, 23, 28, 38, 39, 40, 43, 44, 45, 48, 49, 52, 55, 58, 62, 63, 64, 90, 91, 103, 120
missions, 16, 17, 24, 32, 34, 84
models, 47
momentum, 127
moratorium, 22
motivation, 125
museums, 77

N

naming, 65

National Contingency Plan, 19, 27
National Counterterrorism Center, 56
national emergency, 85, 148
national policy, 41, 59
National Research Council, 66
National Response Framework, viii, 2, 8, 15, 20, 27, 28, 30, 34, 35, 46, 48, 52, 54, 58, 66, 67, 68
National Response Plan, 8, 18, 30, 48
national security, 40, 65, 66
National Security Agency, 67
National Security Council, 49, 65, 66, 69
National Strategy, 59
National Urban Search and Rescue Response System,, x, 72
natural disaster, ix, 7, 15, 17, 24, 26, 28, 37, 38, 39, 42, 46, 50, 65, 89, 93, 117, 118, 129, 135, 140
natural disasters, ix, 7, 24, 28, 37, 38, 39, 42, 46, 50, 65, 89, 93, 118, 129, 135
natural hazards, 7, 42, 51, 57, 106
natural resources, 19, 20
Natural Resources Conservation Service, 22
NCP, 19, 20, 32
New Deal, 117
non-federal resources, ix, 38, 40
non-governmental entities, xii, 145, 146
nonprofit organizations, x, xi, 11, 29, 71, 77, 91, 108, 113, 129, 153
NRF, 8, 15, 16, 18, 24, 27, 28, 30, 54, 55, 58, 59, 67, 68
nursing, 94

O

Obama, ix, 7, 37, 38, 40, 65, 73, 90
Obama Administration, 40, 65, 90
objective criteria, 119, 123, 139
Office of Justice Programs, 149, 150, 151
officials, 11, 20, 23, 27, 28, 39, 47, 49, 73, 75, 81, 83, 88, 92, 95, 96, 98, 99, 101, 105, 109, 110, 121, 125, 126, 133, 135, 137, 152, 153, 158
oil, 7, 8, 15, 19, 20, 21, 27, 33, 38, 43
Oil Pollution Act, 19
Oil Pollution Act (OPA), 19
oil spill, 7, 8, 19, 20, 21, 27, 38, 43
Oil Spill Liability Trust Fund, 20, 33
Oklahoma, 31
OPA, 32
operating costs, 26
operations, 4, 6, 9, 16, 18, 26, 30, 39, 46, 53, 55, 56, 57, 58, 68, 74, 94, 120, 146, 155
opportunities, 28, 59
organize, 17
outreach, 61
overlap, 5, 59
oversight, 8, 18, 20, 97
overtime, 87, 106
ownership, 101

P

participants, 125
Pentagon, 31
permission, iv
permit, 23
petroleum, 32
Philadelphia, 140, 141
plants, 15
police, 30
policy, ix, 13, 28, 30, 31, 34, 38, 39, 40, 41, 42, 43, 45, 47, 48, 49, 54, 55, 56, 57, 58, 63, 64, 65, 66, 79, 97, 99, 102, 109, 125, 135, 139, 158
policy issues, 54
politics, 97
population, 66, 68, 122, 128, 136, 138
population size, 138, 139
portfolio, 56
Posse Comitatus Act, 21, 22, 23
PPD-8 policies, ix, 38, 39, 40, 41, 43, 62, 63, 64

precedent, 39, 54, 93, 140
precedents, 96, 117
preparation, 39, 75, 104
preparedness, viii, ix, x, 2, 3, 4, 5, 6, 7, 18, 24, 28, 30, 34, 38, 39, 40, 41, 42, 43, 44, 45, 47, 48, 49, 53, 54, 57, 61, 63, 64, 69, 72, 73, 81, 134
preservation, 9, 104
president, 97, 123, 126
President Clinton, 141
President Obama, 73, 102
presidential authority, 73
presidential declaration, 9, 21, 74, 97, 115, 122, 126, 130
Presidential Directives, 18
prevention, ix, 7, 30, 34, 38, 40, 44, 49, 50, 52, 56, 57, 63, 91
price changes, 147
price index, 122
primacy, 122
principles, viii, 2, 3, 4, 8
private sector, vii, ix, 1, 3, 8, 38, 41, 61, 65, 91
producers, 22, 157
professionals, 126, 146
profit, ix, 25, 38, 41, 45, 63, 74, 91, 92, 93, 94, 95, 127, 134, 148, 152, 153
program administration, viii, 2
program staff, 129
project, 14, 53, 76, 87, 107
protection, ix, 7, 8, 24, 28, 35, 38, 40, 44, 46, 49, 50, 52, 56, 57, 63, 114, 157
public assistance, 80, 91, 92, 93, 95, 99, 100, 101, 105, 110, 111, 122
public health, xi, 11, 15, 24, 27, 32, 34, 57, 105, 113, 114
Public Health Service Act, 15, 24, 32, 34
public safety, 6, 20, 127, 149
public sector, 124
public service, 23, 79, 93
Puerto Rico, 73

Q

questioning, 95

R

radio, 30
reactions, 121
real estate, 29
real property, 150
recognition, 91
recommendations, xi, 12, 114, 117, 118, 120, 122, 124, 139
reconstruction, 26, 94
recovery process, 11, 60, 146
recreational, 106, 150
recycling, 28
redevelopment, 94
redistribution, 86
reform, 121
Reform, ix, 17, 30, 37, 38, 63, 69, 75, 79, 102, 103, 109, 118, 121
reforms, 17, 121
regulations, 19, 20, 25, 29, 67, 69, 77, 81, 83, 88, 97, 99, 100, 101, 103, 104, 105, 106, 110, 117, 121, 122, 124, 125, 126, 128, 131, 132, 137, 138, 151, 152, 153, 155
regulatory changes, 122
rehabilitation, 29
reimburse, 13, 90, 104, 135
relief, 26, 74, 83, 95, 117, 152, 158
remediation, 20
rent, 85
repair, xii, 6, 11, 26, 29, 31, 76, 77, 79, 87, 90, 91, 92, 107, 131, 146, 147, 150, 153
requirements, ix, x, 7, 14, 17, 18, 25, 28, 30, 37, 39, 40, 45, 46, 49, 53, 59, 60, 62, 65, 66, 72, 74, 75, 76, 81, 82, 86, 87, 93, 97, 98, 101, 102, 106, 111, 115, 116, 122, 149, 151, 152

reserves, 123
Residential, 147, 150
resilience, 39, 44, 46, 49, 51, 57
resolution, 108, 111
resource allocation, 63
responsiveness, 101, 124
restoration, 11, 13, 20, 26, 27, 29, 74, 79, 84, 87, 94
restrictions, 21, 22, 75, 108
retirement, 153
revenue, 74, 77, 90, 154
rights, 34, 83
risk, ix, 7, 13, 30, 37, 39, 43, 46, 47, 49, 51, 57, 58, 66, 138, 157
risk assessment, 47
risk factors, 47
risk management, 157
risks, 5, 6, 46, 51, 84, 129
routes, 24
rules, 5, 25, 101, 110, 148, 158
rural areas, 92

S

SACE, 91
SAFE Port Act, 75
safety, xi, 11, 24, 27, 43, 77, 81, 105, 113, 114
Salvation Army, 12
savings, 129
school, 6, 12, 94, 132
scope, ix, 8, 16, 38, 39, 43, 54, 57, 74, 76, 99, 109, 117, 126, 133
sea-level, 65
sea-level rise, 65
Secretary of Agriculture, 15, 85, 156, 157
Secretary of Commerce, 15
Secretary of Defense, 23, 52, 65, 69, 104
Secretary of Homeland Security, 14, 20, 33, 47, 102
security, ix, 7, 23, 37, 39, 41, 46, 50, 56, 57, 65, 68, 76
security forces, 23
self-employed, 148
Senate, 59, 60, 94, 96, 97, 100, 104, 107, 108, 109, 119, 141
September 11, 18, 31, 93, 94, 126
service provider, 78
services, 6, 11, 12, 17, 23, 26, 76, 77, 78, 82, 84, 85, 89, 91, 93, 95, 106, 108, 151, 152
shape, 54
shelter, xii, 12, 16, 146
small businesses, 95, 156
social welfare, 117
society, 38
solid waste, 28
solution, 122
specialists, 158
spending, xi, 13, 31, 98, 114, 115, 118, 119, 120, 141
Stafford assistance, x, 72, 74, 79
stakeholders, ix, 38, 40, 41, 44, 53, 54, 55, 58, 63
state of emergency, 31
statistics, 136
statutes, 17, 33, 82
statutory authority, 102
statutory provisions, 89
stock, 77, 79
storage, x, 28, 32, 72, 79
storms, 13, 73, 80, 90, 93, 108
stress, 137
structure, 4, 21, 30, 53, 54, 55, 57, 58, 59, 61, 67, 129
subjectivity, 133
Sun, 142
Superfund, 19, 32
supply chain, 56, 68
suppression, 28, 33, 35
surveillance, 6, 7
survivors, x, 25, 52, 66, 68, 72, 74, 98, 149

T

tanks, 32
target, 44, 66
Task Force, 96, 97, 108, 109
tax base, 128
tax credits, 153
taxation, 139
taxes, 33
taxpayers, 134
team members, 78, 126, 138
teams, 6, 55, 78, 81, 125, 126, 137
technical assistance, 24, 27, 28, 74, 146, 154
technologies, 8
technology, 19
telephone, 76
temporary housing, xii, 11, 32, 76, 79, 85, 86, 107, 145, 146, 147
temporary jobs, 149
territorial, xii, 21, 22, 145, 146
territory, 21, 115
terrorism, 7, 18, 19, 38, 39, 42, 45, 46, 47, 50, 51, 52, 55, 56
terrorist acts, ix, 37
terrorist attack, 18, 19, 27, 56, 67, 92, 93, 94, 126, 146
terrorists, 55
testing, 98, 109
theft, 152
threats, ix, 6, 7, 24, 27, 29, 37, 39, 42, 43, 46, 49, 50, 55, 57, 58, 63, 65
time constraints, 110
time frame, 100, 101
Title I, 73, 74, 101, 105, 107, 108, 110, 143
Title II, 73, 74, 105, 108, 143
Title IV, 73, 74, 101, 107, 110
Title V, x, 17, 19, 30, 31, 72, 73, 74, 75, 107
TMC, 94
tornadoes, 119, 130
total costs, 99
tracks, 139
training, 6, 21, 61, 78, 85, 148
transparency, 121
transport, 24
transportation, 7, 23, 25, 26, 46, 50, 57, 66, 68, 76, 85, 87, 104, 138
transportation infrastructure, 26, 66, 68
transportation security, 7, 46, 50, 57
Transportation Security Administration, 67
trauma, 132, 133
Treasury, 139
treatment, 27, 106, 117
trust fund, 20, 26, 33
tuition, 149

U

U.S. Army Corps of Engineers, 24, 91
U.S. Department of Agriculture, 15, 22, 130, 157
U.S. Geological Survey, 142
U.S. Treasury, 33
unemployment insurance, 133
uniform, 40, 117, 140
uninsured, xi, 11, 74, 85, 86, 87, 101, 113, 132, 147, 150, 156
United, vii, viii, x, 1, 2, 3, 9, 10, 11, 15, 18, 19, 23, 28, 31, 38, 39, 40, 67, 71, 105, 107, 108, 116, 123, 142, 153
United States, vii, viii, x, 1, 2, 3, 9, 10, 11, 15, 18, 19, 23, 28, 31, 38, 39, 40, 67, 71, 105, 107, 108, 116, 123, 142, 153
universe, 83
updating, 97
urban, 43, 57
USA, 159
USDA, 22, 33
utility costs, 76

V

variables, 117, 125
variations, 117
vehicles, 150
vein, 133
vessels, 32
Vice President, 139
victims, x, xii, 6, 72, 76, 79, 82, 84, 87, 98, 117, 132, 133, 136, 140, 145, 146, 147, 151
vision, 39, 44, 54
voluntary organizations, 84
vouchers, 25
vulnerability, 49

W

wages, 87, 106
waiver, 13, 31, 35
Washington, 32, 33, 34, 52, 53, 64, 65, 66, 67, 68, 104, 107, 108, 109, 125, 140, 141
waste, 18, 28
waste management, 28
wastewater, 27, 106
water, 6, 10, 22, 24, 27, 35, 67, 73, 82, 106, 132, 153
water resources, 24
water supplies, 27
watershed, 33
weapons, 45, 50, 56
weapons of mass destruction, 45, 50, 56
websites, xii, 146, 159
welfare, 15, 32
wells, 24
White House, 30, 52, 59, 64, 65, 66, 67, 68, 98, 110, 117, 124, 139
wildfire, 28, 35
wildland, 35
wires, 93
withdrawal, 24
witnesses, 110
WMD, 45, 46, 50
workers, 78, 85, 149
workforce, 18, 94
Workforce Investment Act, 149, 158
World Trade Center, 93, 94